Stereochemistry and Bonding

With Contributions by
R. Boča M. Breza A. Ceulemans J. Degen D. M. P. Mingos
M. O'Keeffe P. Pelikán H.-H. Schmidtke
L. G. Vanquickenborne L. Zhenyang

With 50 Figures and 29 Tables

AF736854

Springer-Verlag Berlin Heidelberg GmbH

Editorial Board

Professor *Michael J. Clarke,* Boston College, Department of Chemistry, Chestnut Hill, Massachusetts 02167, U.S.A.

Professor *John B. Goodenough,* Inorganic Chemistry Laboratory, University of Oxford, South Parks Road, Oxford OX1 3QR, Great Britain

Professor *James A. Ibers,* Department of Chemistry, Northwestern University, Evanston, Illinois 60201, U.S.A.

Professor *Christian K. Jørgensen,* Dépt. de Chimie Minérale de l'Université, 30 quai Ernest Ansermet, CH-1211 Genève 4

Professor *David Michael P. Mingos,* University of Oxford, Inorganic Chemistry Laboratory, South Parks Road, Oxford OX1 3QR, Great Britain

Professor *Joe B. Neilands,* Biochemistry Department, University of California, Berkeley, California 94720, U.S.A.

Professor *Graham A. Palmer,* Rice University, Department of Biochemistry, Wiess School of Natural Sciences, P. O. Box 1892, Houston, Texas 77251, U.S.A.

Professor *Dirk Reinen,* Fachbereich Chemie der Philipps-Universität Marburg, Hans-Meerwein-Straße, D-3550 Marburg

Professor *Peter J. Sadler,* Birkbeck College, Department of Chemistry, University of London, London WC1E 7HX, Great Britain

Professor *Raymond Weiss,* Institut Le Bel, Laboratoire de Cristallochimie et de Chimie Structurale, 4, rue Blaise Pascal, F-67070 Strasbourg Cedex

Professor *Robert Joseph P. Williams,* Wadham College, Inorganic Chemistry Laboratory, Oxford OX1 3QR, Great Britain

ISBN 978-3-662-15093-1 ISBN 978-3-540-46083-1 (eBook)
DOI 10.1007/978-3-540-46083-1

Library of Congress Cataloging-in-Publication Data. Stereochemistry and bonding / with contributions by R. Boča ... [et al.]. p. cm. -- (Structure and bonding ; 71) Includes bibliographies and index.

1. Stereochemistry. 2. Chemical bonds. I. Boča, Roman. II. Series. QD461.S92 vol. 71 [QD481] 541.2'2s--dc19 [541.2'23] 89-5913

This work is subject to copyright. All rights are reserved, whether the whole or part of the material is concerned, specifically the rights of translation, reprinting, re-use of illustrations, recitation, broadcasting, reproduction on microfilms or in other ways, and storage in data banks. Duplication of this publication or parts thereof is only permitted under the provisions of the German Copyright Law of September 9, 1965, in its version of June 24, 1985, and a copyright fee must always be paid. Violations fall under the prosecution act of the German Copyright Law.

© Springer-Verlag Berlin Heidelberg 1989
Originally published by Springer-Verlag Berlin Heidelberg New York in 1989.
Softcover reprint of the hardcover 1st edition 1989

The use of general descriptive names, trade marks, etc. in this publication, even if the former are not especially identified, is not to be taken as a sign that such names, as understood by the Trade Marks and Merchandise Marks Act, may accordingly be used freely by anyone.

Typesetting: Mitterweger Werksatz GmbH, 6831 Plankstadt, Germany

2152/3140-543210 – Printed on acid-free paper

71 Structure and Bonding

Editors:
M. J. Clarke, Chestnut Hill · J. B. Goodenough, Oxford
J. A. Ibers, Evanston · C. K. Jørgensen, Genève
D. M. P. Mingos, Oxford · J. B. Neilands, Berkeley
G. A. Palmer, Houston · D. Reinen, Marburg
P. J. Sadler, London · R. Weiss, Strasbourg
R. J. P. Williams, Oxford

Table of Contents

Non-Bonding Orbitals in Co-Ordination, Hydrocarbon and Cluster Compounds
D. M. P. Mingos, L. Zhenyang 1

Vibronic Interactions in the Stereochemistry of Metal Complexes
R. Boča, M. Breza, P. Pelikán 57

A Dynamic Ligand Field Theory for Vibronic Structures Rationalizing Electronic Spectra of Transition Metal Complex Compounds
H.-H. Schmidtke, J. Degen 99

The Epikernel Principle
A. Ceulemans, L. G. Vanquickenborne 125

The Prediction and Interpretation of Bond Lengths in Crystals
M. O'Keeffe 161

Author Index Volumes 1–71 191

Non-Bonding Orbitals in Co-Ordination, Hydrocarbon and Cluster Compounds

D. Michael P. Mingos and Lin Zhenyang

Inorganic Chemistry Laboratory, South Parks Road, Oxford OX1 3QR, U.K.

The origin of non-bonding orbitals in molecular compounds is reviewed and analysed using general quantum mechanical considerations. A combination of the pairing theorem and a group theoretical analysis leads to a definition of the number of non-bonding molecular orbitals in co-ordination, polyene and cluster compounds. The non-bonding molecular orbitals have been generated by defining the nodal characteristics of the relevant orbitals and evaluating the solutions under the appropriate boundary conditions. The stereochemical rôle of non-bonding molecular orbitals in co-ordination compounds is discussed with particular reference to main group and transition metal examples.

1 Introduction 2

2 Non-Bonding Orbitals in Molecular and Co-Ordination Compounds, ML_n 4
2.1 Expansion of Ligand Set as Spherical Harmonic Functions 4
2.1.1 Planar ML_n, Bipyramidal ML_{n+2}, Prismatic and Anti-Prismatic ML_{2n} Complexes 6
2.1.2 Deltahedra 7
2.2 The m − n Rule 8
2.3 The Pairing Theorem 10
2.4 Group Theoretical Aspects 13
2.5 The Effects of Geometric Changes 15
2.6 More Quantitative Analysis of the Effects of *s-p-d* Separation ($\alpha_s \neq \alpha_p \neq \alpha_d$) 24
2.7 Generation of Non-Bonding Orbitals 28
2.8 Summary 30

3 Non Bonding Orbitals in Hydrocarbons 34
3.1 Alternant Hückel Hydrocarbons 34
3.2 Alternant Möbius Hydrocarbons 39

4 Non-Bonding Orbitals in Cluster Compounds 39
4.1 Expansion of the Cluster Skeletal Molecular Orbitals in Terms of the **Tensor Surface Harmonic** Functions 40
4.2 Pairing Principle and Group Theoretical Aspects 41
4.3 Non-Bonding Orbitals in *Closo* Deltahedral and Three-Connected Cluster Compounds 42
4.4 Non-Bonding Orbitals in *Nido* Structures 44

5 Summary 54

6 References 55

© Springer-Verlag Berlin Heidelberg 1989

1 Introduction

Phrases such as "lone pairs", "empty non-bonding orbitals", "stereochemically active (or non-active) lone pairs", and "electron-rich molecules" are part of the language of modern chemistry and their frequent use is a testimony to their usefulness in chemical communication. In common with electronegativity they are very difficult to define precisely and this imprecision can at times lead to heated debate. In this review we shall not attempt to define these terms rigorously, but to codify the situations where "non-bonding orbitals" occur and attempt to provide some general principles for understanding the stereochemical and chemical consequences of these concepts.

Historically, the concept of non-bonding electron pairs can be traced to Lewis' classic work on the definition of a covalent bond in terms of sharing electrons between atoms [1]. The electrons not involved in such bonds were considered independently and described as 'lone pairs'. He recognized that although such electron pairs did not contribute at all to the bonding within the molecule, they could be donated to a second molecule with an incomplete octet and thereby form a dative bond. These ideas were given a quantum mechanical interpretation in Pauling's Valence Bond Theory [2]. He showed how molecules could be built up by using the concept of hybridisation. Linear combinations of atomic orbitals that were directed towards the vertices of regular polyhedra maximised the overlap between orbitals on the central atom and the orbitals at the vertices [3]. If each of these orbitals contained a single electron then an electron pair bond resulted. 'Lone pairs' in molecules such as ammonia (**1**) also utilised hybrid orbitals which had been constructed for the tetrahedron and were populated by an electron pair from the nitrogen. The non-bonding nature of this orbital was assured, because it was

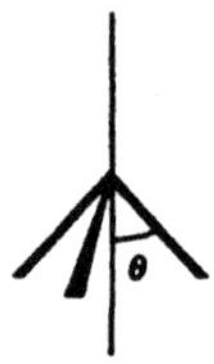

(1)

defined as being orthogonal to the other sp^3 hybrid orbitals of the NH_3 molecule. Within the framework of valence bond theory it is also possible to have non-bonding vacant orbitals based either on hybrids or on unhybridised orbitals that have nodal planes coincident with the positions of the peripheral atoms of the molecule. For example, the p_z orbital in planar BX_3 (**2**) has a nodal plane coincident with the positions of the X atoms. Similarly the d_{xz}, d_{yz} and d_{xy} orbitals of an octahedral molecule are non-bonding

(2)

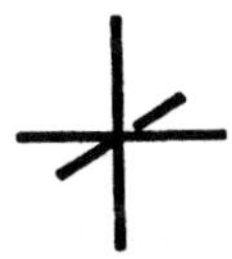

(3)

because they have nodal planes that are coincident with the octahedral bond directions **(3)**. The relative energetics of utilising an unhybridised orbital in preference to a hybridised one was not clearly made in the initial valence bond formulations. Furthermore, the deviations from the idealized hybridization angles required that the lone pair contain a different proportion of *s* character than the idealized hybrids. The geometric and electronic consequences of these differences in hybridization were effectively developed by Bent [4].

The **Valence Shell Electron Pair Theory** [5–7] returned to the classical Lewis description of the chemical bond, but introduced a three-dimensional aspect to it by calculating the preferred polyhedra for n electron pairs using a point charge repulsion model. This, when combined with some simple rules defining the relative importance of "lone-pair – lone-pair" and "lone-pair – bond pair" repulsions, provides a pedagogically useful way of accounting for the geometries of main group molecules. It does, however, have serious limitations when applied to transition metal complexes.

In the molecular orbital method [8] the bonding is described in terms of linear combinations of atomic orbitals and the localised description of the chemical bond is lost, but as Mulliken [9] showed, it is possible to partition the electron density and thereby get an estimate of the spatial distribution and bonding character of an orbital. The procedure can be illustrated by the simple case of a diatomic molecule with one basis function per atom containing N electrons in a molecular orbital. Let the molecular orbital be written as:

$$\phi = a\phi_A + b\phi_B$$

where ϕ_A and ϕ_B are atomic orbitals of atoms A and B. The total population of the MO is:

$$Na^2\int\phi_A^2 d\tau + Nb^2\int\phi_B^2 d\tau + 2\,Nab\int\phi_A\phi_B d\tau$$

Mülliken described Na^2 the net electron population in ϕ_A on A, Nb^2 that in ϕ_B on B and $2\,Nab\langle\phi_A|\phi_B\rangle$ the overlap population, since $\int\phi_A\phi_B d\tau$ occupies the region of overlap of the two atomic orbitals.

A chemist also wants to know the gross populations on each atom. The gross population on A or B certainly includes the respective net population, but for each atom one must also include a part of the overlap population. The simplest procedure is to proportion the two parts equally to A and B, even when a and b are unequal i.e., $Nab\langle\phi_A|\phi_B\rangle$ to each atom. Adding this to the net population of either atom gives a gross population N(A) or N(B), such that N(A) + N(B) = N. Therefore, a specific molecular orbital can be described as nonbonding either because N(A) $\approx$ N i.e., it is localised almost exclusively on one centre, or N(A) and N(B) are both significant but the overlap population $2\,Nab\langle\phi_A|\phi_B\rangle$ is approximately equal to zero.

In this review we will describe the bonding in main group and transition metal compounds in terms of a molecular orbital framework and attempt to define more precisely those situations where the conditions for non-bonding character are met. Such a discussion will focus primarily on the nodal characteristics of the molecular orbitals, since the overlap populations between atoms in the non-bonding character approach zero and the analysis of the overlap populations reflect this fact. Such a definition is useful for discussing the stereochemical role of orbitals within an isolated molecule, but does not provide a guide to the ability of that orbital to interact with orbitals on other molecules. This is influenced by the ionisation energy of the non-bonding orbital and its radial characteristics. We are mindful of this and will make passing references to the availability of 'non-bonding orbitals' for intermolecular bond formation. Our initial discussion will focus on the bonding in spherical co-ordination compounds where it is convenient to express the linear combinations of atomic orbitals as a spherical harmonic expansion.

2 Non-Bonding Orbitals in Molecular and Co-Ordination Compounds, ML_n

In the majority of co-ordination and molecular compounds the number of ligands is not equal to the number of valence orbitals on M and it is necessary to establish the factors that influence the resultant pattern of molecular orbitals. It will be demonstrated below that ML_n is generally characterized by m-n non-bonding molecular orbitals where m is the number of the valence orbitals in the central atom. Before the discussion of this "m-n rule", the symmetry adapted linear combinations of ligand σ orbitals are discussed.

2.1 *Expansion of the Ligand Set as Spherical Harmonic Functions*

In the Complementary Spherical Electron Density Model [10–12], the symmetry adapted linear combinations of σ_i ligand orbitals for ML_n compound can be expressed in terms of a spherical harmonic expansion:

$$\Psi_{lm} = N' \sum_i C_{l,m}(\theta_i, \phi_i)\sigma_i$$

$$= L_m^\sigma, \qquad m = 0, 1\,c, 1\,s, \ldots \qquad L = S, P, D, \ldots,$$

where θ_i and ϕ_i represent the locations of the ligand nuclei in spherical polar coordinates, N' is a normalizing constant, and $C_{l,m}$ is a modified spherical harmonic wave function, $C_{l,m} = [4\,\pi/(2\,l + 1)]^{1/2} Y_{l,m}$. The modified spherical harmonic functions are given in Table 1. For those functions with imaginary solutions the real solutions are obtained by taking linear combinations of complementary components i.e.,

$$C_{l,m}^c = 1/(2)^{1/2}[(-1)^m C_{l,m} + C_{l,-m}]$$

$$C_{l,m}^s = 1/i(2)^{1/2}[(-1)^m C_{l,m} - C_{l,-m}]$$

Table 1. Polar forms of the modified spherical harmonic functions

C_{lm}	Polar form
C_{00}	1
C_{10}	$\cos\theta$
C_{11}^c	$\sin\theta\cos\phi$
C_{11}^s	$\sin\theta\sin\phi$
C_{20}	$1/2\,(3\cos^2\theta - 1)$
C_{21}^c	$(3)^{1/2}\cos\theta\sin\theta\cos\phi$
C_{21}^s	$(3)^{1/2}\cos\theta\sin\theta\sin\phi$
C_{22}^c	$(3/4)^{1/2}\sin^2\theta\cos 2\phi$
C_{22}^s	$(3/4)^{1/2}\sin^2\theta\sin 2\phi$
C_{30}	$1/2\,(5\cos^3\theta - 3\cos\theta)$
C_{31}^c	$(3/8)^{1/2}\sin\theta(5\cos^2\theta - 1)\cos\phi$
C_{31}^s	$(3/8)^{1/2}\sin\theta(5\cos^2\theta - 1)\sin\phi$
C_{32}^c	$(15/4)^{1/2}\cos\theta\sin^2\theta\cos 2\phi$
C_{32}^s	$(15/4)^{1/2}\cos\theta\sin^2\theta\sin 2\phi$
C_{33}^c	$(5/8)^{1/2}\sin^3\theta\cos 3\phi$
C_{33}^s	$(5/8)^{1/2}\sin^3\theta\sin 3\phi$

In this fashion the linear combinations are assigned quantum numbers l and m which are related to those which have been defined for the S, P and D functions derived for the particle on the sphere problem. Furthermore, their nodal characteristics mimic those of the atomic wave functions of the central atom. The spherical harmonic expansion described above will provide its most accurate description of the symmetry adapted linear combinations when the polyhedral vertices are symmetry equivalent. For example, octahedral ML_6 has S^σ, $P^\sigma_{0,\pm1}$ and $D^\sigma_{0,2s}$ a total of 6 symmetry adapted linear combinations. They are

$$S^\sigma = 1/(6)^{1/2}(\sigma_1 + \sigma_2 + \sigma_3 + \sigma_4 + \sigma_5 + \sigma_6)$$

$$P^\sigma_0 = 1/(2)^{1/2}(\sigma_1 - \sigma_6)$$

$$P^\sigma_{1c} = 1/(2)^{1/2}(\sigma_2 + \sigma_4)$$

$$P^\sigma_{1s} = 1/(2)^{1/2}(\sigma_3 - \sigma_5)$$

$$D^\sigma_0 = 1/(3)^{1/2}(\sigma_1 - 1/2\,\sigma_2 - 1/2\,\sigma_3 - 1/2\,\sigma_4 - 1/2\,\sigma_5 + \sigma_6)\ ;$$

$$D^\sigma_{2c} = 1/2(\sigma_2 - \sigma_3 + \sigma_4 - \sigma_5)\ .$$

They are shown schematically in Fig. 1. This spherical expansion improves as an approximation as n increases. When there is more than one linear combination with the same symmetry then an orthogonalisation problem can arise which will be discussed in more detail below. In the following sections the utilisation of these spherical harmonic expansions are summarised for a range of common ligand co-ordination geometries.

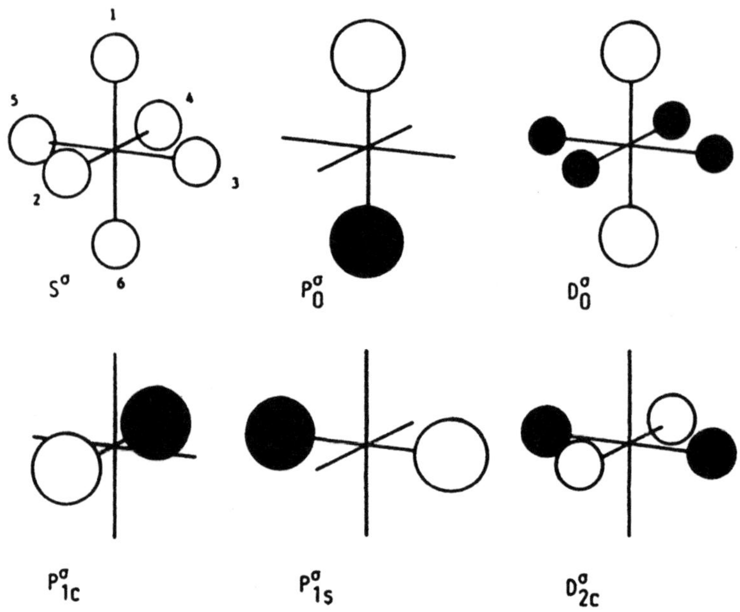

Fig. 1. The linear combinations of ligand σ orbitals in an octahedral compound

2.1.1 *Planar ML_n, Bipyramidal ML_{n+2}, Prismatic and Anti-Prismatic ML_{2n} Complexes*

In the planar structure, ML_n, if all the ligands are defined as lying in the xy plane, those functions that possess a nodal plane coincident with the xy plane are systematically excluded. Therefore, the ligand orbitals span the S^σ, $P^\sigma_{\pm 1}$, $D^\sigma_{\pm 2}$, ..., $L^\sigma_{\pm L}$, ... spherical harmonic expansions and there are total of n wave functions. For example, the L_3 moiety has S^σ, and $P^\sigma_{\pm 1}$ wave functions. For L_4 the additional function generated is a D function i.e., it is characterised by S^σ, $P^\sigma_{\pm 1}$ and D^σ_{2s}.

In the bipyramidal (ML_{n+2}) compounds, the two additional vertices which are located in the + z and − z directions generate two more L^σ's i.e., P^σ_0 and D^σ_0, besides the n L^σ's generated by the S^σ, $P^\sigma_{\pm 1}$, $D^\sigma_{\pm 2}$, ..., $L^\sigma_{\pm L}$, ... series for the planar geometry which belong to different irreducible representations. They are all orthogonalised with the exception of the S^σ and D^σ_0. The re-orthogonalisation of S^σ and D^σ_0 can be achieved by assuming $\Psi_1 = S^\sigma$ and $\Psi_2 = aS^\sigma + bD^\sigma_0$. The a and b parameters can be obtained by solving the following equations

$$\begin{aligned} \langle\Psi_1|\Psi_2\rangle &= 0 \\ \langle\Psi_2|\Psi_2\rangle &= 1\,. \end{aligned} \quad \longrightarrow \quad \begin{aligned} a + b\langle S^\sigma|D^\sigma\rangle &= 0 \\ a^2 + b^2 + 2\,ab\langle S^\sigma|D^\sigma_0\rangle &= 1 \end{aligned}$$

For example, the σ ligand orbitals in a trigonal bipyramid ML_5 span S^σ, $P^\sigma_{0,\pm 1}$ and D^σ_0 functions. S^σ and D^σ_0 are

$$S^\sigma = 1/(5)^{1/2}(\sigma_1 + \sigma_2 + \sigma_3 + \sigma_4 + \sigma_5)$$

$$D_0^\sigma = 1/(11)^{1/2}(2\,\sigma_1 - \sigma_2 - \sigma_3 - \sigma_4 + 2\,\sigma_5)$$

where the two axial vertices are defined as positions 1 and 5.

$$\langle S^\sigma | D_0^\sigma \rangle = 1/(55)^{1/2}(2 - 1 - 1 - 1 + 2) = 1/(55)^{1/2}$$

Therefore, $a = 1/(54)^{1/2}$ and $b = -(55/54)^{1/2}$ are obtained by solving the above equations. It is noteworthy that the spherical approximation is a reasonable one even in such cases.

In the prismatic structures ML_{2n}, the 2n ligands are taken as the combinations of two n-planar moieties. The L^σ's in ML_{2n} are the linear combinations of the two subsets in the two planes i.e., $\Psi^\sigma_{plane1} \pm \Psi^\sigma_{plane2}$. Since the Ψ^σ_{plane1} and Ψ^σ_{plane2} span the same series of S^σ, $P^\sigma_{\pm1}$, $D^\sigma_{\pm2}$, ..., $L^\sigma_{\pm L}$, ..., one of the combinations i.e., $\Psi^\sigma_{plane1} + \Psi^\sigma_{plane2}$, has the same characteristics as that of the individual plane, Ψ^σ_{plane1} or Ψ^σ_{plane2}. The other, $\Psi^\sigma_{plane1} - \Psi^\sigma_{plane2}$, generates one more horizontal nodal plane between the two planes. Therefore, $\Psi^\sigma_{plane1} - \Psi^\sigma_{plane2}$ is obtained by the multiplication rule developed by Johnston and Mingos [13].

$$\{S^\sigma, P^\sigma_{\pm1}, D^\sigma_{\pm2}, \ldots, L^\sigma_{\pm L}, \ldots\} \otimes P^\sigma_0$$

$$\rightarrow \{P^\sigma_0, D^\sigma_{\pm1}, F^\sigma_{\pm2}, \ldots, (L+1)^\sigma_{\pm L}, \ldots\}$$

For example, for the pentagonal prism, ML_{10}, the ligand orbitals span S^σ, $P^\sigma_{\pm1}$ and $D^\sigma_{\pm2}$ which are the same as the L^σ's for pentagonal planar ML_5, and P^σ_0, $D^\sigma_{\pm1}$, $F^\sigma_{\pm2}$ which are generated by the above multiplication.

The anti-prismatic structures ML_{2n} are constructed from two n-planar moieties which are staggered with respect to each other. For odd n, all of the L^σ functions generated by an anti-prism are the same as those generated by the corresponding prism because the L^σ's of the two n-planar moieties span the same series of L^σ's. The situation for even n, however, is different. In this case, the expansions of the L^σ wave functions for the upper and lower n-planar moieties are different in the final term. If the L^σ's of the upper n-planar moiety span the S^σ, $P^\sigma_{\pm1}$, $D^\sigma_{\pm2}$, ..., $(\Lambda - 1)^\sigma_{\pm(\Lambda-1)}$, Λ^σ_c (where $\Lambda = L_{max} = n/2$), the L^σ's of the lower one span the S^σ, $P^\sigma_{\pm1}$, $D^\sigma_{\pm2}$, ..., $(\Lambda - 1)^\sigma_{\pm(\Lambda-1)}$, Λ^σ_s. Therefore, all the L^σ's for anti-prisms when n is even are:

$$S^\sigma, P^\sigma_{\pm1}, D^\sigma_{\pm2}, \ldots, (\Lambda - 1)^\sigma_{\pm(\Lambda-1)}$$

and

$$\{S^\sigma, P^\sigma_{\pm1}, D^\sigma_{\pm2}, \ldots, (\Lambda - 1)^\sigma_{\pm(\Lambda-1)}\} \otimes P^\sigma_0 \text{ plus } \Lambda^\sigma_c \text{ and } \Lambda^\sigma_s.$$

2.1.2 Deltahedra

In a deltahedral ML_n compound, the linear combinations of the n ligand orbitals are built up in such a way that the n lowest L quantum states are always used before higher L quantum states. For example, ML_4 with a tetrahedral arrangement of four ligands generates S^σ and three P^σ wave functions, and an octahedral ML_6 has S^σ, three P^σ and two D^σ wave functions. Unlike the previous classes of polyhedra, F^σ functions are not used prior to the completion of the D shell. This property can be related to the ability of deltahedra to give the best coverage on a spherical surface [11].

The spherical harmonic expansion methodology developed above provides a method of defining non-bonding orbitals that are localized exclusively on the ligand atoms. In particular, if the ligands have an L^σ function which is not matched by a valence orbital on the central atom in terms of quantum numbers or point group symmetry, then it is strictly non-bonding. For example, in an ML_8 co-ordination compound, which has a cubic arrangement of ligands, the L^σ ligand linear combinations are S^σ, $P^\sigma_{\pm 1}$, D^σ_{2s}, P^σ_0, $D^\sigma_{\pm 1}$ and F^σ_{2c}. If M is a transition metal atom the *s, p* and *d* valence orbitals cannot match the F^σ_{2s} function which in the O_h point group transforms as a_{2u}. Consequently there is a non-bonding orbital which is localised exclusively on the ligands because of its symmetry and pseudo-spherical symmetry characteristics. The situation is slightly less well defined if the L^σ linear combination does not match the pseudo-spherical symmetry requirements but has the same symmetry as one of the valence orbitals of the central atom. For example, a first-row main group atom which has *s* and *p* valence orbitals might form a trigonal bipyramidal complex in a transition state. The relevant L^σ linear combinations are S^σ, $P^\sigma_{0,\pm 1}$ and D^σ_0. The latter transforms as a_1' in the D_{3h} point group i.e., the same irreducible representation as *s*. Therefore, some overlap between *s* and D^σ_0 can occur, but it is likely to be small because the D^σ_0 and *s* functions have different pseudo-symmetries. In particular the D^σ_0 function is doubly noded and the *s* function is un-noded. Pseudo-symmetry arguments of this type are very important for simplifying the bonding problems in co-ordination compounds.

2.2 The *m* − *n* Rule

In an n-coordinate compound, we assume that there are in total m atomic orbitals (*s, p, d, ...*) (m = 4 or 9 for great majority of situations) associated with the central atom. The new linear combinations of the n σ_i ligand orbitals, $\Psi^\sigma_1, \Psi^\sigma_2, \ldots, \Psi^\sigma_n$, which are normalised and orthogonal to each other (i.e., $\langle \Psi^\sigma_i | \Psi^\sigma_j \rangle = \delta_{ij}$), can always be derived from group theoretical methods and the spherical harmonic classification illustrated above. If the Coulomb integrals for the *m* atomic orbitals of the central atom are assumed to have the same α_c, the corresponding secular determinant in the Hückel approximation is:

$$\begin{array}{c} \\ n \\ \\ m \end{array} \begin{vmatrix} \overbrace{\begin{matrix} \alpha_l - E & & \\ & \ddots & \\ & & \alpha_l - E \end{matrix}}^{n} & \overbrace{\mathbf{R}}^{m} \\ \mathbf{R}^t & \begin{matrix} \alpha_c - E & & \\ & \ddots & \\ & & \alpha_c - E \end{matrix} \end{vmatrix} = 0 \qquad (1)$$

where α_l is the Coulomb integral for the ligand σ orbitals and **R** is the n × m matrix representing the interactions between n ligands and the central atomic orbitals ($\mathbf{R}^t$ is the transpose of **R**). Solving Eq. (1), we get m + n eigenvalues. In mathematical terms this corresponds to obtaining the eigenvalues of matrix A:

$$\begin{pmatrix} \alpha_l & & & & & \\ & \ddots & & & \mathbf{R} & \\ & & \alpha_l & & & \\ & & & \alpha_c & & \\ & \mathbf{R}^t & & & \ddots & \\ & & & & & \alpha_c \end{pmatrix} \tag{2}$$

$|A - EI| = 0$, I is the identity matrix. Changing the form of (2), we get

$$A = \begin{pmatrix} \alpha_l - \alpha_c & & & \\ & \ddots & & \mathbf{R} \\ & & \alpha_l - \alpha_c & \\ & \mathbf{R}^t & & \mathbf{O} \end{pmatrix} \begin{matrix} n \\ \\ m \end{matrix} \quad \begin{matrix} \\ n \qquad m \end{matrix} \tag{3}$$

where **O** is an (m × m) zero matrix. The relationship between the eigenvalues of (3) and (2) is

$$E' = E + \alpha_c \tag{4}$$

E′ and E are the eigenvalues of Eqs. (2) and (3), respectively.

The number of zero eigenvalues of Eq. (3) (i.e., the number of non-bonding orbitals) is equal to or larger than (m − n) when m ≥ n.

Proof: The more general form of Eq. (3) is:

$$A = \begin{pmatrix} \mathbf{B}_{n,n} & \mathbf{C}_{n,m} \\ \mathbf{D}_{m,n} & \mathbf{O}_{m,m} \end{pmatrix} \tag{5}$$

where $\mathbf{O}_{m,m}$ is an (m × m) zero matrix.

According to the expansion of the characteristic equation [14],

$$\det(A - EI) = E^{(n+m)} - G_1(A)E^{(n+m)-1} + G_2(A)E^{(n+m)-2} - \ldots \pm G_{(n+m)}(A) \tag{6}$$

where the $G_k(A)$ is the sum of the k-by-k principal minors of matrix A. The k-by-k principal minor is the determinant of k-by-k principal submatrix. A k-by-k principal submatrix of A is one lying in the same set of k rows and columns in A ($k \le n + m$). For example, the largest k-by-k principal minor $G_{(n+m)}(A)$ is det(A).

All the k-by-k principal minors of matrix (5), when $k > 2n + 1$, are zero because any one of the k-by-k principal minors can be rearranged in the following form:

$$B_k = \begin{pmatrix} B_1 & B_2 \\ B_3 & O_{(k-n,k-n)} \end{pmatrix} \quad (7)$$

where $O_{(k-n,k-n)}$ is a zero matrix.

The expansion of the determinant of B_k is

$$\det(B_k) = \sum_{\sigma} \text{sign}(\sigma) \prod_{i=1}^{k} b_{i\sigma(i)} ,$$

where $b_{i\sigma(i)}$ is the element of the matrix B_k. The sum runs over all k permutations of the k items {1, 2, ..., k} and the sign is + 1 or − 1 according to whether the minimum number of transpositions necessary to achieve it starting from {1, 2, ..., k} is even or odd. Each product in the summation is of the form

$$b_{1\sigma(1)}b_{2\sigma(2)} \ldots b_{k\sigma(k)} \quad (8)$$

If $k \geq 2n + 1$, there is at least one zero element involved in Eq. (8). Thus, $\det(B_k) = 0$. Therefore, $G_k(A) = 0$ as well when $k \geq 2n + 1$. Now,

$$\begin{aligned} &\det(A - EI) \\ &= E^{(n+m)} - G_1(A)E^{(n+m)-1} + G_2(A)E^{(n+m)-2} - \ldots \pm G_{2m}(A)E^{(n+m)-2n} \\ &= E^{(m-n)}[E^{2n} - G_1(A)E^{2n-1} + G_2(A)E^{2n-2} - \ldots \pm G_{2n}(A)] \end{aligned}$$

So the multiplicity of $E = 0$ eigenvalues is eqaul to or larger than $(m - n)$. From (4), we say that there are at least $(m - n)$ orbitals with the energy of α_c which are localised on the central atom.

Similarly, if $n > m$, we get at least $(n - m)$ orbitals with the energy of α_l which are localised on the ligands i.e., $n - m$ non-bonding orbitals.

The following generalisation can be derived when the proof above is combined with the arrangement of the ligand shell in ML_n. *The number of non-bonding orbitals in ML_n systems is exactly equal to m − n if the n ligands define a three-dimensional arrangement.* This is because the ligand orbitals span the S^σ, P^σ, D^σ, ..., functions, giving a total of n wave functions while the D^σ functions are not used prior to the completion of the P^σ functions. Therefore, there is a one-to-one pseudo-symmetry matching between *s*, *p* (and *d*) orbitals on the central atom and the ligand orbitals, leaving $m - n$ redundant wave functions. *The most important implication from the m − n rule is that the number of non-bonding orbitals is independent of the geometry of ML_n if the arrangement of the ligand shell retains three or two dimensions. A change in geometry only leads to a change in the character of the non-bonding orbitals i.e., the ratio of s, p and d character.*

2.3 The Pairing Theorem

When $\alpha_c = \alpha_l = \alpha$, the pairing theorem which has been widely used in discussion of alternant conjugated organic molecules [15–19] is applicable to the system above because the magnitude of each element in **R** of Eq. (2) does not affect the generalisation of this proof. This theorem states that if $\alpha + E_k$ is an eigenvalue of Eq. (2), $\alpha - E_k$ is also an eigenvalue of Eq. (2). Therefore, the molecular orbitals of such a system are perfectly paired in the Hückel approximation. The corresponding paired molecular orbitals are:

$$\Psi_k^+ = \sum_i c_{ik}\phi_i + \sum_l \sum_m c_{lm(k)} L_m^\sigma \tag{9}$$

$$\Psi_k^- = \sum_i c_{ik}\phi_i - \sum_l \sum_m c_{lm(k)} L_m^\sigma \tag{10}$$

where ϕ_i are the valence atomic orbitals on the central atom and c_{ik} and $c_{lm(k)}$ are the coefficients of ϕ_i and L_m^σ in the kth molecular orbital, respectively. More precisely, the approximation $\langle L_m^\sigma|\phi_{lm'}\rangle = 0$ (unless L = l and m = m') leads to a diagonalised secular matrix which can be separated into sets of 2 by 2 matrices in which the ϕ_{lm} and L_m^σ are involved. The resonance integral $\langle L_m^\sigma|H|\phi_{lm}\rangle = \beta_{lm}$ for a specific value of l and m leads to pairs of molecular orbitals:

$$\Psi_+ = \phi_{lm} + L_m^\sigma \tag{11 a}$$

$$\Psi_- = \phi_{lm} - L_m^\sigma \tag{11 b}$$

with orbital energies of $\alpha + \beta_{lm}$ and $\alpha - \beta_{lm}$ corresponding the in-phase and out-of-phase combinations of the central atom ϕ_{lm} and ligand L_m^σ orbitals. Equations (11 a) and (11 b) are simplifications of Eqs. (9) and (10).

The octahedral ML_6 is an example where the approximation is perfect. The resonance integrals, $\langle L_m^\sigma|H|\phi_{lm}\rangle$, can be obtained by using the following equations:

$$\Psi_{l,m}^\sigma = N' \sum_i^n C_{l,m}(\theta_i, \phi_i)\sigma_i\ ;$$

$$\begin{aligned}\langle \Psi_{l,m}^\sigma|\phi_{l,m}\rangle &= N'\left\langle \sum_i^n C_{l,m}(\theta_i, \phi_i)\sigma_i|\phi_{l,m}\right\rangle ;\\ &= N' \sum_i^n C_{l,m}(\theta_i, \phi_i)\langle\sigma_i|\phi_{l,m}\rangle\\ &= N' \sum_i^n C_{l,m}(\theta_i, \phi_i) \times C_{l,m}(\theta_i, \phi_i)S_\sigma(l)\end{aligned}$$

(where $S_\sigma(l)$ is defined in (4))

$$\begin{aligned}&= N' \sum_i^n C_{l,m}^2(\theta_i, \phi_i)S_\sigma(l)\\ &= \left[\sum_i^n C_{l,m}^2(\theta_i, \phi_i)\right]^{1/2} S_\sigma(l)\end{aligned}$$

$$\left(\text{because } N' = 1\Big/\left[\sum_i^n C_{l,m}^2(\theta_i, \phi_i)\right]^{1/2}\right).$$

The overlap integrals between the L^σ's and atomic orbitals of central atom in ML_6 octahedral structure are:

$$\langle S^{\sigma}|s\rangle = (6)^{1/2}S_{\sigma}(s)$$
$$\langle P^{\sigma}_{m}|p_{m}\rangle = (2)^{1/2}S_{\sigma}(p) \quad m = 0, \pm 1$$
$$\langle D^{\sigma}_{m}|d_{m}\rangle = (3)^{1/2}S_{\sigma}(d) \quad m = 0, 2c\,.$$

and

$$\langle D^{\sigma}_{m}|d_{m}\rangle = 0 \qquad m = \pm 1, 2s$$

Therefore,

$$\langle L^{\sigma}_{m}|H|\phi_{l,m}\rangle = k\langle L^{\sigma}_{m}|\phi_{l,m}\rangle^{2}$$
$$= \sum_{i} C^{2}_{l,m}(\theta_i, \phi_i)kS^{2}_{\sigma}(l)$$
$$= \sum_{i} C^{2}_{l,m}(\theta_i, \phi_i)\beta_{\sigma}(l)$$

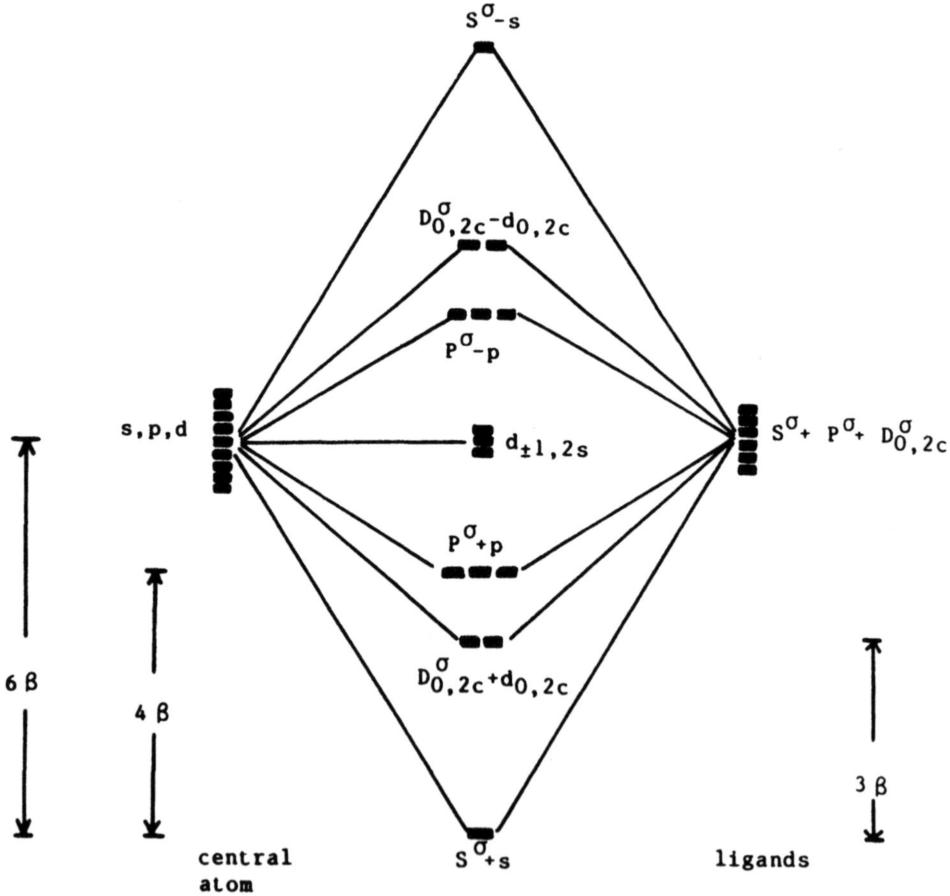

Fig. 2. The orbital interaction for an octahedral ML_6 compound

These lead to the following results:

$\langle S^{\sigma}|H|s\rangle = 6\ \beta_{\sigma}(s)$

$\langle P^{\sigma}_{m}|H|p_{m}\rangle = 2\ \beta_{\sigma}(p) \quad m = 0, \pm 1$

$\langle D^{\sigma}_{m}|H|d_{m}\rangle = 3\ \beta_{\sigma}(d) \quad m = 0, 2\,c$

and

$\langle D^{\sigma}_{m}|H|d_{m}\rangle = 0 \quad m = \pm 1, 2\,s$

It is noteworthy that the stabilisation energies are always $N\beta^{\sigma}$/Degeneracy where N = Number of ligand atoms. When the assumption $\beta_{\sigma}(s) = \beta_{\sigma}(p) = \beta_{\sigma}(d)$ has been made, the orbital interaction diagram for this idealized situation ($\alpha_s = \alpha_p = \alpha_d = \alpha_l$) is shown in Fig. 2.

2.4 Group Theoretical Aspects

A conjugate organic system is described as alternant if the atoms in it can be divided into two sets *starred* and *unstarred* in such a way that no two atoms in the same set are directly linked. In the Hückel approximation, this definition implies that all the p_{π} atomic orbitals in the *starred* or *unstarred* set are orthogonal to each other (i.e., no interaction occurs between the p_{π}'s in the same group) and there are interactions between the two groups. Therefore, the secular determinant describing the orbital interactions of an alternant system have the same form as Eq. (1), except α_c and α_l are set to be α and the resonance integrals are β or 0. Consequently, the orbital interactions in an ML_n compound can be described in a similar fashion to those of an alternant hydrocarbon if the m valence orbitals on the central atom are *starred* and the ligand orbitals are *unstarred* because the valence orbitals on the central atom are orthogonal to each other and have zero resonance integrals among them. The ligand orbitals behave likewise.

In a discussion of the group theoretical aspects of alternant systems [20] given in Sect. 3, we have defined an alternant operator, $\hat{A}$, which interconverts the *starred* and *unstarred* atomic positions. Since the central atom must lie on the principal axis the alternant operator belongs to the totally symmetric representation in the ML_n system. Therefore, the paired molecular orbitals $\Psi_k^{|}$ (9) and Ψ_k (10) belong to the same irreducible representations which are called self-conjugate representations i.e., the in-phase (bonding) and the corresponding out-of-phase (antibonding) combinations must belong to the same irreducible representations: the bonding and antibonding combinations differ by the addition of a radial rather than an angular node.

Group theory can be used to resolve the secular determinant (1) into a block-diagonalised matrix. Each block matrix represents the interactions between the basis sets belonging to the same irreducible representation.

When the arguments developed above are combined with the **pairing theorem** and the **m − n rule,** the following important generalization which applies to all simple mononuclear compounds ML_n can be derived.

The non-bonding molecular orbitals in ML_n compounds always arise from the presence of an odd number of irreducible representations with the same symmetry. The number of such non-bonding orbitals is equal to the sum of the degeneracies of the irreducible representations which occur in odd numbers. For example, if one system has odd numbers of a and e irreducible representations, the number of non-bonding orbitals is 1 + 2. *ML_n compounds where the number of valence electron pairs is m have $|m - n|$ non-bonding molecular orbitals.*

The occurrence of these non-bonding orbitals can be associated with either spherical or non-spherical co-ordination polyhedra. The choice of co-ordination polyhedra depends on several factors and notably:
(a) The relative energies of the valence orbitals
(b) The population of the non-bonding orbitals
(c) Multiple bonding effects
(d) Steric effects.
The relative importance of these factors are described in more detail in the subsequent Sections. The group theoretical conclusion above can be illustrated by the following examples.

Simple co-ordination and main group molecules with no non-bonding molecular orbitals (i.e., n = m), e.g., $[ReH_9]^{2-}$ [21] and CH_4, have pairs of self-conjugate bonding and antibonding molecular orbitals, e.g. a_1, t_2, a_1^*, t_2^* for CH_4. When n − m = 1 e.g., BH_3, NH_3, $[H_4Mo(PPh_3)_4]$ (dodecahedron), $[H_4W(CN)_8]$ (square antiprism) [22] there is one non-bonding molecular orbital resulting from an odd number of self-conjugate representations i.e., $a_2''(BH_3)$, $a_1(NH_3)$, $b_1[H_4Mo(PPh_3)_4]$, $a_1[H_4W(CN)_8]$. When n − m = 2 e.g., BeH_2, OH_2, $[Mo(CNBu^t)_7]^{2+}$ (capped trigonal prism), $[V(CN)_7]^{4-}$ (pentagonal bipyramid), and $W(CO)_4Br_3$ (capped octahedron) [22] there must either be a odd number of self-conjugate *e* representations or a pair of orbitals with different symmetries each of which arises from an odd number self-conjugate representations. For example, BeH_2 (π_u, e_{1u}), OH_2 (a_1 and b_2), $[Mo(CNBu^t)_7]^{2+}$ (a_1, b_2), $[V(CN)_7]^{4-}$ (e_1''), and $W(CO)_4Br_3$ *(e)*.

Table 2. Examples of co-ordination compounds with variable occupations of the non-bonding orbitals

Example	Co-ordination geometry (number)	d^n
$TiCl_4(diars)_2$	dodecahedron (8)	0
$NbCl_4(diars)_2$	dodecahedron (8)	1
$MoH_4(PH_3)_4$	dodecahedron (8)	2
$[ZrF_7]^{2-}$	pentagonal bipyramid (7)	0
$[VF_7]^{4-}$	pentagonal bipyramid (7)	2
$OsH_4(PMe_2Ph)_3$	pentagonal bipyramid (7)	4
$[TiF_6]^{2-}$	octahedron (6)	0
OsF_6	octahedron (6)	2
PtF_6	octahedron (6)	4
$Cr(CO)_6$	octahedron (6)	6

When $n - m = 3$ e.g., $Mo(CO)_6$ (octahedral), $Mo(S_2C_2H_2)_3$ (trigonal prism) the three non-bonding orbitals must be either an odd number of self-conjugate t representations, or e and a_1 representations or three singly degenerate representations with different symmetries e.g., $Mo(CO)_6$ (t_{2g}), $Mo(S_2C_2H_2)_3$ (e' and a_1'). The full utilisation of the bonding molecular orbitals and the non-bonding molecular orbitals defined above leads naturally to the inert gas rule for co-ordination and organometallic compounds. The non-bonding nature of these orbitals in transition metal compounds is confirmed by the partial occupation of these orbitals. Some specific examples of complexes illustrating this point are summarised in Table 2. Clearly the arguments can be extended to hypervalent compounds where $n > m$ and there are $n - m$ non-bonding orbitals localised on the ligand sphere. Such compounds exceed the inert gas rule by $2(m - n)$ valence electrons. For example, in PF_5 and SF_6 there are $1(a_1)$ and $2(e_g)$ non-bonding L^σ orbitals, respectively, localised on the fluorine ligands.

2.5 The Effects of Geometric Changes

In this section the effect of a geometrical distortion on the character and energies of the non-bonding molecular orbitals will be considered by reference to specific examples. In the main group, molecule MH_3 (**2**) with a planar D_{3h} geometry, the $m - n$ rule suggests the occurrence of 1 non-bonding orbital and this can be readily be identified as p_z (see (**2**)). If the molecule is distorted to give a pyramidal C_{3v} geometry (**1**) then the s and p_z orbitals belong to the same a_1 irreducible representation of the C_{3v} point group. The energies of the resultant a_1 molecular orbitals are obtained by solving the following determinant:

$$\begin{array}{cccc} s & p_z & \Psi^\sigma & \\ A = \begin{vmatrix} \alpha - E & 0 & \beta_s \\ 0 & \alpha - E & \beta_p \\ \beta_s & \beta_p & \alpha - E \end{vmatrix} & & & \begin{array}{c} s \\ p_z \\ \Psi^\sigma \end{array} \end{array}$$

$$\beta_s = \langle \Psi^\sigma | H | s \rangle$$

$$\beta_p = \langle \Psi^\sigma | H | p_z \rangle$$

$$\Psi^\sigma = 1/(3)^{1/2}(\sigma_1 + \sigma_2 + \sigma_3)$$

$$\det(A) = (\alpha - E)^3 - (\alpha - E)\beta_s^2 - (\alpha - E)\beta_p^2 = 0$$

This determinant leads of course to a non-bonding orbital with energy, α, confirming the $(m - n)$ rule, but the wave function of this molecular orbital now has the following form:

$$\Psi_{n\cdot b} = \Psi(2\,a_1) = \beta_p/(\beta_s^2 + \beta_p^2)^{1/2}\, s - \beta_s/(\beta_s^2 + \beta_p^2)^{1/2}\, p_z$$

This non-bonding orbital is illustrated in Fig. 3 for angles $\theta = 90° - 54.73°$. It is noteworthy that the overlap integral between $\Psi_{n\cdot b}$ and the ligand orbital Ψ^σ is zero i.e.,

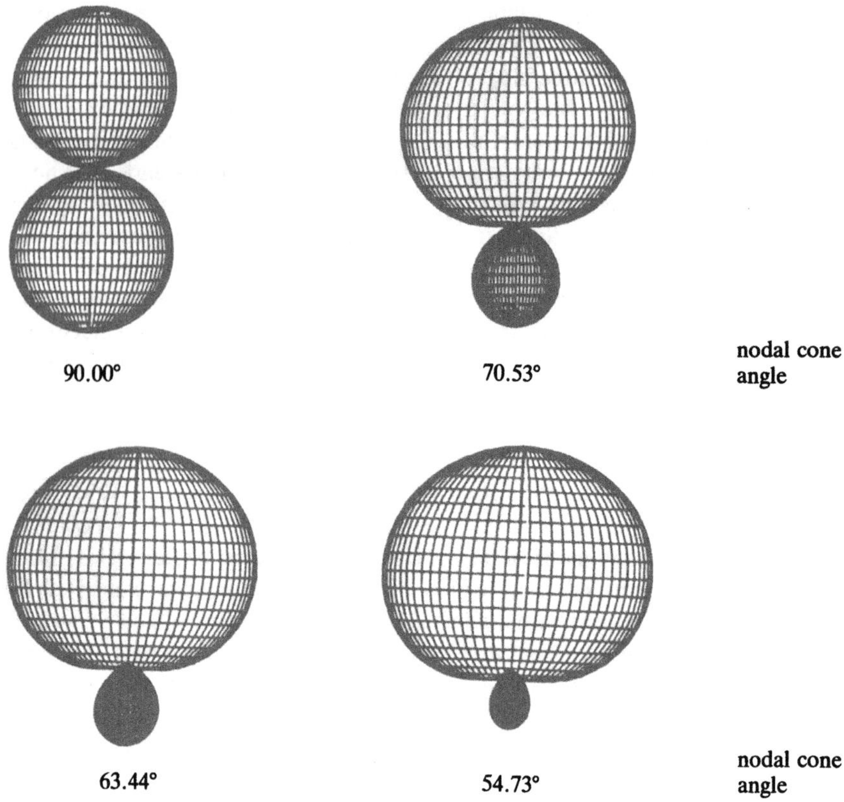

Fig. 3. The a_1 non-bonding orbital for MH_3 (**1**) with different θ angles

$$\langle \Psi_{n\text{-}b} | \Psi^{\sigma} \rangle = 0 \tag{12}$$

Clearly, as the hydrogen atoms move away from the original MH_3 plane, the s and p orbitals mix to produce a nodal cone coincident with the M–H bond directions. In this way the non-bonding character is maintained. The computed s and p characters as a function of θ are summarised in Table 3 for the situation where

$$\beta_\sigma(s) = \beta_\sigma(p) \qquad \text{(as defined in (4))}$$

$$\beta_s = \langle 1/(3)^{1/2}(\sigma_1 + \sigma_2 + \sigma_3)|H|s\rangle = (3)^{1/2}\beta_\sigma(s)$$

$$\beta_p = \langle 1/(3)^{1/2}(\sigma_1 + \sigma_2 + \sigma_3)|H|p_z\rangle = -(3)^{1/2}\cos\theta\,\beta_\sigma(p)\,.$$

(4)

Table 3. The bonding and non-bonding molecular orbitals in MH_3 as a function of θ angle for the situation where $\beta_\sigma(s) = \beta_\sigma(p)$

θ (Degree) [Bond angle]	Symmetry	Bonding and non-bonding molecular orbitals
90.00°	a''_2:	p_z
[120.00°]	1 e':	$0.707\,p_x + 0.707\,P^{\sigma}_{x}$, $0.707\,p_y + 0.707\,P^{\sigma}_{y}$
	a'_1:	$0.707\,s + 0.707\,S^{\sigma}$
70.53°	2 a_1:	$0.316\,s + 0.949\,p_z$
[109.47°]	1 e:	$0.707\,p_x + 0.707\,P^{\sigma}_{x}$, $0.707\,p_y + 0.707\,P^{\sigma}_{y}$
	1 a_1:	$0.685\,s - 0.224\,p_z + 0.685\,S^{\sigma} - 0.224\,P^{\sigma}_{0}$
68.00°	2 a_1:	$0.351\,s + 0.936\,p_z$
[106.80°]	1 e:	$0.707\,p_x + 0.707\,P^{\sigma}_{x}$, $0.707\,p_y + 0.707\,P^{\sigma}_{y}$
	1 a_1:	$0.662\,s - 0.248\,p_z + 0.662\,S^{\sigma} - 0.248\,P^{\sigma}_{0}$
63.44°	2 a_1:	$0.408\,s + 0.913\,p_z$
[101.54°]	1 e:	$0.707\,p_x + 0.707\,P^{\sigma}_{x}$, $0.707\,p_y + 0.707\,P^{\sigma}_{y}$
	1 a_1:	$0.645\,s - 0.299\,p_z + 0.645\,S^{\sigma} - 0.299\,P^{\sigma}_{0}$
54.73°	2 a_1:	$0.500\,s + 0.866\,p_z$
[90.00°]	1 e:	$0.707\,p_x + 0.707\,P^{\sigma}_{x}$, $0.707\,p_y + 0.707\,P^{\sigma}_{y}$
	1 a_1:	$0.612\,s - 0.354\,p_z + 0.612\,S^{\sigma} - 0.354\,P^{\sigma}_{0}$

The alternant nature of the non-bonding molecular orbital problem as represented in (5) means that the coefficients at the starred positions multiplied by the relevant resonance integrals must add up to zero at the unstarred position:

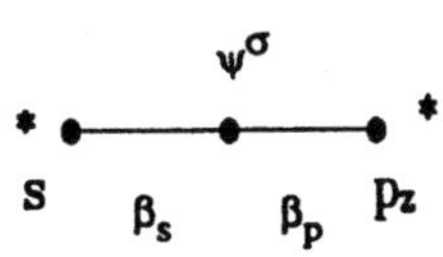

(5)

$$\beta_s C_s + \beta_p C_p = 0$$

i.e.,

$$C_s/C_p = -\beta_p/\beta_s$$

Specifically for this problem:

$$C_s/C_p = [\beta_\sigma(p)/\beta_\sigma(s)]\cos\theta = \cos\theta$$

This result conforms with the computed coefficients given in Table 3.

The energies of the bonding and antibonding molecular orbitals can also be evaluated from the secular determinant. They are:

$$E(1\,a_1) = \alpha - (\beta_s^2 + \beta_p^2)^{1/2}$$

$$E(3\,a_1) = \alpha + (\beta_s^2 + \beta_p^2)^{1/2}$$

and the relevant wave functions are:

$$\Psi(1\,a_1) = 1/(2)^{1/2}[1/(\beta_s^2 + \beta_p^2)^{1/2}(\beta_s s + \beta_p p_z)] + 1/(2)^{1/2}\Psi^\sigma$$

$$\Psi(3\,a_1) = 1/(2)^{1/2}[1/(\beta_s^2 + \beta_p^2)^{1/2}(\beta_s s + \beta_p p_z)] - 1/(2)^{1/2}\Psi^\sigma$$

The energies and wave functions emphasise the alternant nature of the problem. If we express Ψ^σ in terms of a linear combination of S^σ and P_z^σ weighted according to their overlap integrals:

$$\Psi^\sigma \propto \langle\Psi^\sigma|s\rangle S^\sigma + \langle\Psi^\sigma|p_z\rangle P_z^\sigma$$

$$= 1/(\beta_s^2 + \beta_p^2)^{1/2}[(\beta_s S^\sigma + \beta_p P_z^\sigma)]$$

Therefore,

$$\Psi(1\,a_1) = 1/(2)^{1/2}[1/(\beta_s^2 + \beta_p^2)^{1/2}(\beta_s s + \beta_p p_z)] + 1/(2)^{1/2}[1/(\beta_s^2 + \beta_p^2)^{1/2}(\beta_s S^\sigma + \beta_p P_z^\sigma)]$$

$$\Psi(2\,a_1) = 1/(\beta_s^2 + \beta_p^2)^{1/2}(\beta_p s - \beta_s p_z)$$

From these equations it is apparent that the sum of the squares of the coefficients associated with s, S^σ, p_z and P_z^σ for all the bonding and non-bonding orbitals have following relationships:

$$C_s^2 + C_{S^\sigma}^2 = 1$$

$$C_p^2 + C_{P^\sigma}^2 = 1$$

These relationships are confirmed in Table 3.

This represents a justification of the Spherical Electron Density Model [11] which suggests that in molecules that conform to the inert gas rule the total electron density is pseudo-spherical.

The corresponding molecular orbitals derived from p_x and p_y have the following energies and wave functions:

$$E_{\text{bonding}}(1\,e) = \alpha + \beta(p_{x,y})$$

$$\beta(p_{x,y}) = \langle p_x|H|P_x^\sigma\rangle = \langle p_y|H|P_y^\sigma\rangle = (6)^{1/2}/2 \sin\theta\, \beta_\sigma(p)$$

$$1\,e_x = 1/(2)^{1/2}(p_x + P_x^\sigma) \qquad P_x^\sigma = 1/(6)^{1/2}[2\,\sigma_1 - (\sigma_2 + \sigma_3)]$$

$$1\,e_y = 1/(2)^{1/2}(p_y + P_y^\sigma) \qquad P_y^\sigma = 1/(2)^{1/2}(\sigma_2 - \sigma_3)$$

$C_p^2 + C_{P^\sigma}^2 = 1$ for 1 e_x and 1 e_y bonding orbitals.

In Fig. 4 the energies of the molecular orbitals as a function of θ are illustrated for the situation where $\alpha = 0$, $\beta_\sigma(s) = \beta_\sigma(p) = \beta$ and represent a "Walsh diagram" for the

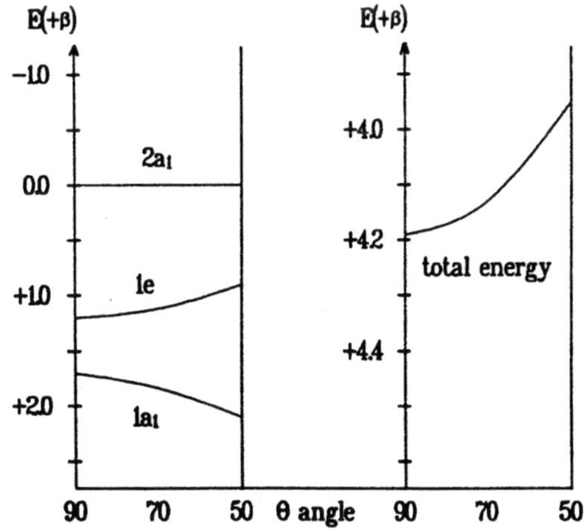

Fig. 4. The energies of molecular orbitals in MH_3 (**1**) as a function of θ angle

distortion co-ordinate. The validity of the pairing theorem along the distortion co-ordinate is nicely illustrated in the diagram. It is significant that, except for the situation where the lowest energy bonding molecular orbital only is filled, the planar geometry is preferred for all remaining electron counts up to the inert gas configuration. The more symmetrical geometry is preferred because the destabilisation energies associated with 1 e bonding orbitals are larger than the stabilisation energy associated with 1 a_1.

The overlap integrals between the hydrogen atoms and the s orbitals of the central atom are independent of the θ angle and therefore cannot contribute to a geometric preference. The p_x and p_y overlap integrals decrease by $\sin\theta$ as a result of the change in geometry and the destabilisation which results outweighs the increased stabilisation associated with the p_z orbital whose overlap integral increases by a factor of $\cos\theta$. Therefore, for the hypothetical situation when $\alpha_s = \alpha_p$ a spherical co-ordination geometry is preferred because it maximises the overlaps between the ligand orbitals and the central atomic valence orbitals.

Interestingly, the geometric change causes a redistribution of electron density between the s and p_z orbitals, which belong to the same a_1 representation of the C_{nv} point

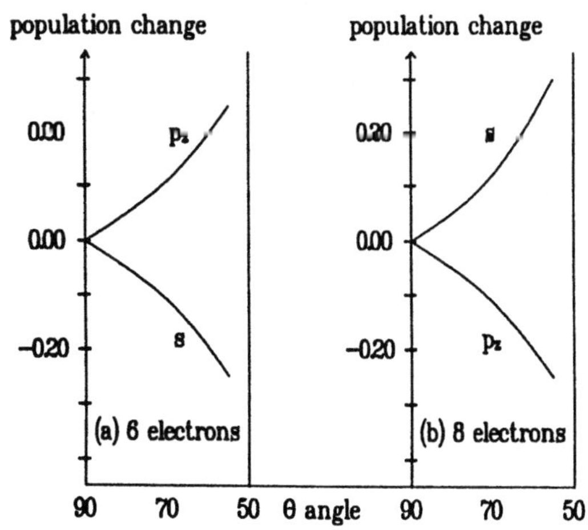

Fig. 5. The redistribution in MH_3 (**1**) molecules of the electron density associated with s and p_z orbitals for 6 electrons (**a**) and 8 electrons (**b**) as a function of θ angle

group. The electron density associated with the p_x and p_y orbitals remains constant, however, because they belong to a different *(e)* irreducible representation. The redistribution of electron density associated with *s* and p_z is illustrated in Figs. 5a and 5b for 6 and 8 electron pyramidal molecules. For the 6 electron molecule the initial populations when $\theta = 90°$ are $s^1p_z^0$ and change to $s^{0.75}p_z^{0.25}$ when $\theta = 54.73°$. In contrast, for the 8 electron molecule the populations change from $s^1p_z^2$ ($\theta = 90°$) to $s^{1.25}p_z^{1.75}$ when $\theta = 54.73°$. This is a very important result because it suggests that a geometric distortion which results in a low symmetry molecule belonging to a C_{nv} point group *causes a transfer of electron density from s to p_z if an empty non-bonding orbital is created and a transfer of electron density from p_z to s if a filled non-bonding hybrid orbital is created.* This has no energetic consequences if $\alpha_s = \alpha_p$, however, in real main group molecules $\alpha_s \neq \alpha_p$ and this transfer of electron density has important geometric consequences. In particular, if the valence *s* orbitals are more stable than the valence *p* orbitals this means that the higher symmetry geometry is preferred unless the non-bonding orbital is occupied. Therefore, the important stereochemical activity of lone pairs in main group chemistry is intimately connected with the redistribution of electron density between *s* and p_z and the relative ordering of the valence orbitals. This aspect will be discussed in a more quantitative fashion below.

The a_1 non-bonding orbital in a square pyramidal ML_5 (**6**) behaves in an analogous manner. The wave function of this a_1 non-bonding orbital can also be obtained in the

(6)

Hückel approximations. In the square pyramid, the 5 L^σ's are $P^\sigma_{\pm 1}$, D^σ_{2c} (which are the same as in the octahedron) and two a_1 representations,

$$\Psi_2 = (1/6)^{1/2}S^\sigma - (1/2)^{1/2}P^\sigma_0 + (1/3)^{1/2}D^\sigma_0 \text{ (localised on position 6)}$$

$$\Psi_3 = (2/3)^{1/2}S^\sigma - (1/3)^{1/2}D^\sigma_0 \text{ (localised on the square)}$$

where S^σ, P^σ_0 and D^σ_0 represent the ligand L^σ functions of an octahedron. These two a_1 ligand orbitals interact with the *s*, p_z and d_{z^2} orbitals of the central atom.

We assume that the resonance integrals $\beta_\sigma(s)$, $\beta_\sigma(p)$ and $\beta_\sigma(d)$ (see (**4**)) are the same and equal to unity. The latter assumption does not affect the coefficients of the molecular orbitals because they define only the orbital energies. Using all the assumptions above and the resonance integrals in Sect. 2.3, the resonance integrals between *s*, p_z and d_{z^2} on the central atom and ligand orbitals Ψ_2 and Ψ_3 can be calculated and are listed in the Table below.

$\langle\Psi_i\|H\|\phi_i\rangle$	s	p_z	d_{z^2}
Ψ_2	1	−1	1
Ψ_3	2	0	−1

Let $H_{ii}(s) = H_{ii}(p) = H_{ii}(d) = \alpha_c$ (c: central atom) and H_{ii}(ligand) $= \alpha_l$ (l: ligand). Then the secular determinant is,

$$\begin{vmatrix} \alpha_l - E & 0 & 1 & -1 & 1 \\ 0 & \alpha_l - E & 2 & 0 & -1 \\ 1 & 2 & \alpha_c - E & 0 & 0 \\ -1 & 0 & 0 & \alpha_c - E & 0 \\ 1 & -1 & 0 & 0 & \alpha_c - E \end{vmatrix} = 0$$

Obviously, there is one eigenvalue with $E = \alpha_c$. This means this orbital with α_c orbital energy is the non-bonding orbital. The non-bonding orbital is:

$$\Psi_{nb} = 1/(14)^{1/2} s + 3/(14)^{1/2} p_z + 2/(14)^{1/2} d_{z^2}$$

localised on the central atom with a maximum density in the direction of the missing vertex (position 1). It can be checked that

$$\langle \Psi_{nb} | \Psi_2 \rangle = \langle \Psi_{nb} | \Psi_3 \rangle = 0 \tag{13}$$

The corresponding bonding molecular orbitals of a_1 symmetry are:

$$\phi_1 = -0.070\, s + 0.406\, p_z - 0.575\, d_{z^2} + \text{ligand contributions}$$

$$E = \alpha + 1.608\, \beta$$

$$\phi_2 = 0.678\, s - 0.116\, p_z - 0.164\, d_{z^2} + \text{ligand contributions}$$

$$E = \alpha + 2.327\, \beta$$

In addition, there are the following molecular orbitals of b_2 and e symmetry:

$$\phi(b_2) = 0.707\, d_{x^2-y^2} + 0.707\, D^{\sigma}_{2c} \qquad E = \alpha + 1.7321\, \beta$$

$$\phi(e_x) = 0.707\, p_x + 0.707\, P^{\sigma}_x$$

$$E = \alpha + 1.414\, \beta$$

$$\phi(e_y) = 0.707\, p_y + 0.707\, P^{\sigma}_y$$

The occupation of the five bonding molecular orbitals leads to the following orbital occupations:

$s^{0.92}p^{2.36}d^{1.72}$

When the non-bonding orbital is occupied these orbital populations become:

$s^{1.06}p^{3.64}d^{2.29}$

For a square pyramidal ML_5 complex, where M is a transition metal, there are, in addition, three non-bonding orbitals, d_{xy}, d_{xz}, d_{yz} localized exclusively on the central

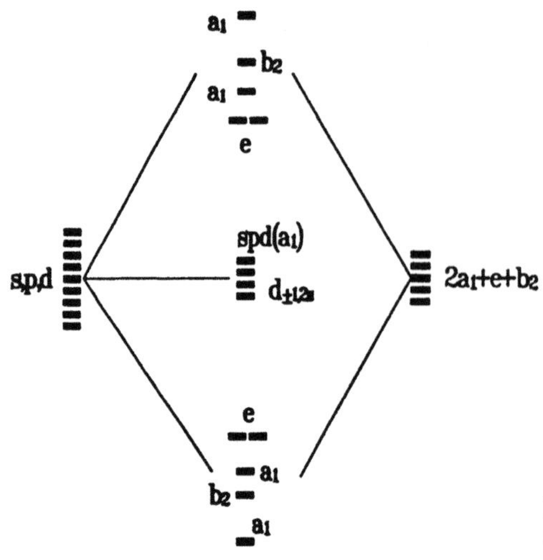

Fig. 6. The orbital interaction diagram for a square pyramidal transition metal ML_5 complex

metal atom. The complete molecular orbital splitting diagram for such a complex is illustrated in Fig. 6.

In a transition metal trigonal bipyramidal complex the corresponding five bonding molecular orbitals are:

$$\phi(1\,a_1') = 0.707\,s + 0.707\,S^\sigma \qquad E = \alpha + 2.236\,\beta$$

$$\phi(2\,a_1') = 0.707\,d_{z^2} + 0.707\,D_0^\sigma \qquad E = \alpha + 1.658\,\beta$$

$$\phi(a_2'') = 0.707\,p_z + 0.707\,P_0^\sigma \qquad E = \alpha + 1.414\,\beta$$

$$\phi(e_x') = 0.707\,[(4/7)^{1/2}p_x + (3/7)^{1/2}d_{x^2-y^2}] + 0.707\,P_x^\sigma$$

$$\phi(e_y') = 0.707\,[-(4/7)^{1/2}p_y + (3/7)^{1/2}d_{xy}] + 0.707\,P_y^\sigma \qquad E = \alpha + 1.620\,\beta$$

In addition, there are the following non-bonding orbitals:

$$\phi_{nb}(e_x') = -(3/7)^{1/2}p_x + (4/7)^{1/2}d_{x^2-y^2}$$

$$E = \alpha$$

$$\phi_{nb}(e_y') = (3/7)^{1/2}p_y + (4/7)^{1/2}d_{xy}$$

and pure d orbitals d_{xz} and d_{yz}. It is significant that in the D_{3h} point group that s and p_z belong to different irreducible representations and therefore cannot mix to create a non-bonding a_1 molecular orbital. However, (p_x, p_y), and $(d_{x^2-y^2}, d_{xy})$ both transform as e' and create linear combinations which either have maxima in the equatorial bond directions leading to $\phi(e_x')$ and $\phi(e_y')$ above or are noded in these directions $\phi_{nb}(e_x')$ and $\phi_{nb}(e_y')$. The relevant molecular orbital diagram is illustrated in Fig. 7.

If only the bonding electron pairs are occupied, the relevant orbital populations are:

$$s^{1.00}d^{1.86}p^{2.14}$$

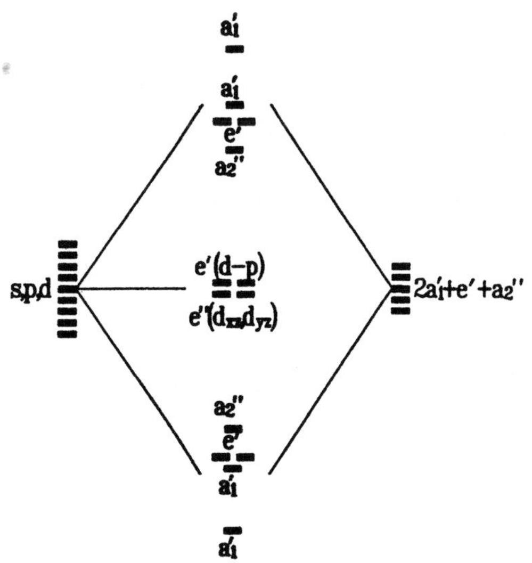

Fig. 7. The orbital interaction for a trigonal bipyramidal transition metal ML_5 complex

If the bonding orbitals are all doubly occupied by electrons, the corresponding total energies are identical for the trigonal bipyramid and square pyramid i.e., 8.55 β. This is consistent with the observation that the potential energy surface for the interconversion of trigonal bipyramidal and square-pyramidal molecules is very soft [23].

The orbital populations are also very similar i.e., $s^{1.00}d^{1.86}p^{2.14}$ for the trigonal bipyramid and $s^{0.92}d^{1.72}p^{2.36}$ for the square pyramid. Therefore, for molecules where the assumption $\alpha_s \neq \alpha_p \neq \alpha_d$ is not valid, there will be little change in the relative stabilities of the two geometries.

The observed atomic energies for transition metals $|\alpha_d| > |\alpha_s| \gg |\alpha_p|$ will lead to a slight preference for the trigonal bipyramidal geometry. For main group atoms where $|\alpha_s| > |\alpha_p| \gg |\alpha_d|$ there will be a corresponding slight preference for the trigonal bipyramid which has a somewhat larger s orbital population.

When the ML_5 molecules have more than five electron pairs the geometric preferences depend on the number of electron pairs and the relative energies of the valence orbitals. For a transition metal where $|\alpha_d| > |\alpha_s| \gg |\alpha_p|$ the following conclusions are pertinent:

a) $d^0 - d^4$(low spin) the two geometries will have similar energies.
b) d^6 (low spin) the square pyramid is energetically preferred because the six electrons can be accommodated in pure d orbitals, whereas for a trigonal bipyramid there is an electron pair in a $d - p$ hybrid.
c) d^8 (low spin) The structures will be finely balanced because for both geometry there are electrons occupying hybrids which have higher orbital energies than pure d orbitals i.e., for a square-pyramid there is one electron pair in an spd hybrid and for a trigonal bipyramid two electron pairs in $p - d$ hybrids.

In a square-planar geometry, the Ψ_1 and Ψ_2 are localised on positions 1 and 6 respectively, where $\Psi_1 = (1/6)^{1/2}S^\sigma + (1/2)^{1/2}P_0^\sigma + (1/3)^{1/2}D_0^\sigma$. The remaining Ψ_3 should be taken into account. Now the secular determinant is:

$$\begin{vmatrix} \alpha_1 - E & 2 & 0 & -1 \\ 2 & \alpha_c - E & 0 & 0 \\ 0 & 0 & \alpha_c - E & 0 \\ -1 & 0 & 0 & \alpha_c - E \end{vmatrix} = 0$$

There are two eigenvalues $E_1 = E_2 = \alpha_c$ i.e., two non-bonding orbitals. They are:

$$\Psi_1 = p_z$$

$$\Psi_2 = 1/(5)^{1/2} s + 2/(5)^{1/2} d_{z^2}$$

localised on the central atom. The overlap integrals between these two non-bonding orbitals and ligand orbital Ψ_3 are zero i.e.,

$$\langle \phi_1 | \Psi_3 \rangle = \langle \phi_2 | \Psi_3 \rangle = 0 \, . \tag{14}$$

The discussion presented above illustrates the form of the non-bonding orbitals in main group compounds and transition metal complexes. Particularly striking is the fact that **all the non-bonding orbitals localised on the central atom have zero overlaps with all the ligands.**

2.6 More Quantitative Analysis of the Effects of s-p-d Separations ($\alpha_s \neq \alpha_p \neq \alpha_d$)

In molecules such as ammonia the energy of the 2 p orbital of nitrogen is similar to that of the hydrogen 1 s orbital but the 2 s orbital is considerably more stable (by ca. 10.00 ev). A first-order correction [24, 25] to the wave functions arising from a change in Coulomb integral from α to $\alpha + \delta\alpha$ is given by:

$$\Psi_i' = \Psi_i^0 + \sum_j t_{ji}^{(1)} \Psi_j^0$$

$$t_{ij} = (C_{\alpha i}^0 C_{\alpha j}^0 \delta\alpha)/(e_i^0 - e_j^0)$$

where Ψ_j^0 and e_j^0 are the zeroth order orbital and energy, and t_{ij} is the mixing coefficient.

For ammonia which has a bond angle of 106.8° and $\theta = 68.0°$, the relevant wave functions of a_1 symmetry are:

	Energy
$\Psi^0(3\,a_1) = 0.662\,s - 0.248\,p_z - 0.707\,\Psi^\sigma(a_1)$	$\alpha + (\beta_s^2 + \beta_p^2)^{1/2}$
$\Psi^0(2\,a_1) = 0.351\,s + 0.936\,p_z$	α
$\Psi^0(1\,a_1) = 0.662\,s - 0.248\,p_z + 0.707\,\Psi^\sigma(a_1)$	$\alpha - (\beta_s^2 + \beta_p^2)^{1/2}$

consider the 1 a_1 orbital:

$$t_{2a1,1a1} = 0.662 \times 0.351 \times [-\delta\alpha/(\beta_s^2 + \beta_p^2)^{1/2}]$$

$$t_{3a1,1a1} = 0.662 \times 0.662 \times 0.5 \times [-\delta\alpha/(\beta_s^2 + \beta_p^2)^{1/2}]$$

Making the simplifying assumption that

$$\delta\alpha = -(\beta_s^2 + \beta_p^2)^{1/2}$$

$$t_{2a1,1a1} = 0.232$$

$$t_{3a1,1a1} = 0.219$$

$$\Psi'(1\,a_1) = 1/N[(1\,a_1) + t_{2a1,1a1}(2\,a_1) + t_{3a1,1a1}(3\,a_1)]$$
$$= 0.803\,s - 0.077\,p_z + 0.591\,\Psi^{\sigma}(a_1)$$

Under these conditions the corresponding wave functions for 3 a_1 is:

$$\Psi'(3\,a_1) = 1/N[(1\,a_1) - t_{2a1,1a1}(2\,a_1) - t_{3a1,1a1}(3\,a_1)]$$
$$= 0.415\,s - 0.391\,p_z - 0.821\,\Psi^{\sigma}(a_1)$$

For the non-bonding 2 a_1 orbital the corresponding perturbation theory correction is:

$$t_{1a1,2a1} = -0.232 \qquad t_{3a1,2a1} = 0.232$$

$$\Psi'(2\,a_1) = 0.333\,s + 0.889\,p_z - 0.312\,\Psi^{\sigma}(a_1)$$

It can be seen that the 1st order perturbation correction leads to the mixing of ligand character into the non-bonding orbital (2 a_1). No change in the C_s/C_p ratio in the modified wave function (2 a_1) occurs because $t_{1a1,2a1} = -t_{3a1,2a1}$. The change of the s and p characters is a result of the re-normalisation. Therefore, *the nodal cone angle in the non-bonding orbital does not change even when mixing of ligand character occurs.* Since there is no s contribution to the e orbitals, the energies and wave functions of the e orbitals are unchanged.

These modified wave functions together with those derived from ab initio calculations [26] are summarised in Table 4. The agreement between the two is remarkably good although the time of computation is several orders of magnitude different.

The first-order correction to the orbital energies is given by:

$$\delta E = (C^0_{\alpha i})^2 \times \delta\alpha$$

A more accurate orbital energies can be calculated from the modified wave functions $\Psi'(1\,a_1)$, $\Psi'(2\,a_1)$ and $\Psi'(3\,a_1)$:

$$E = \langle \Psi'(na_1)|H|\Psi'(na_1)\rangle$$

These are computed with $\alpha = -14.00$ ev, $\delta\alpha = -10.00$ ev, $\beta_s = -9.35$ ev and $\beta_p = 3.50$ ev (because $\delta\alpha = -(\beta_s^2 + \beta_p^2)^{1/2}$ and $\beta_s = (3)^{1/2}\beta_\sigma$, $\beta_p = -(3)^{1/2}\cos\theta\,\beta_\sigma$) and contrasted with the ab initio results [26] in Table 4. The agreement is less good. This

Table 4. The wave functions and orbital energies for NH_3 derived from our approximation and ab initio calculations

ab initio[a]: Orbital		Energy (ev)	Wave function
	$2a_1$	−11.22	$0.30\,s + 0.94\,p_z - 0.07\,[H_{1s}(1) + H_{1s}(2) + H_{1s}(3)]$
	$1e$	−16.97	$0.80\,p_y + 0.43\,[H_{1s}(2) - H_{1s}(3)]$
	$1a_1$	−31.16	$0.87\,s - 0.18\,p_z + 0.19\,[H_{1s}(1) + H_{1s}(2) + H_{1s}(3)]$

this approximation:

Orbital	Energy (ev)		1st order modified wave function
	1st order correction	from modified mos	
$2a_1$	−17.51	−17.33	$0.33\,s + 0.89\,p_z - 0.18\,[H_{1s}(1) + H_{1s}(2) + H_{1s}(3)]$
$1e$	−20.13	−20.13	$0.71\,p_y + 0.50\,[H_{1s}(2) - H_{1s}(3)]$
$1a_1$	−30.62	−26.63	$0.80\,s - 0.08\,p_z + 0.34\,[H_{1s}(1) + H_{1s}(2) + H_{1s}(3)]$

[a] Only 2 *s* and 2 *p* atomic orbitals are considered

emphasises the fact that semi-quantitative molecular orbital methods like the Hückel and extended Hückel do yield reasonably accurate wave functions but the energies of the molecular orbitals are less reliable.

The same methodology can be applied to H_2O which has a bond angle of 104.5°, $\theta = 52.25°$. When $\alpha_s = \alpha_p = \alpha_H$

$$C_s/C_p = -\beta_\sigma(p)\cos\theta/\beta_\sigma(s)$$

for the non-bonding orbital ($2a_1$). Using $\beta_\sigma(p) = \beta_\sigma(s)$

Table 5. The wave functions and orbital energies for H_2O derived from our approximation and ab initio calculations

ab initio[a]: Orbital		Energy (ev)	Wave function
	$1b_1$	−10.95	p_y
	$2a_1$	−12.68	$0.52\,s + 0.79\,p_z - 0.26\,[H_{1s}(1) + H_{1s}(2)]$
	$1b_2$	−16.97	$0.62\,p_x + 0.42\,[H_{1s}(1) - H_{1s}(2)]$
	$1a_1$	−34.95	$0.84\,s - 0.13\,p_z + 0.15\,[H_{1s}(1) + H_{1s}(2)]$

this approximation:

Orbital	Energy (ev)		1st order modified wave function
	1st order correction	from modified mos	
$1b_1$	−14.00	−14.00	p_y
$2a_1$	−21.83	−21.17	$0.48\,s + 0.78\,p_z - 0.24\,[H_{1s}(1) + H_{1s}(2)]$
$1b_2$	−24.12	−24.12	$0.71\,p_x + 0.50\,[H_{1s}(1) - H_{1s}(2)]$
$1a_1$	−38.00	−32.48	$0.79\,s - 0.14\,p_z + 0.42\,[H_{1s}(1) + H_{1s}(2)]$

[a] Only 2 *s* and 2 *p* atomic orbitals are considered

$\Psi_{nb}(2\ a_1) = 0.522\ s + 0.853\ p_z$

The resultant wave functions derived from first-order perturbation theory with $\delta\alpha = -(\beta_s^2 + \beta_p^2)^{1/2} = -15.00$ ev ($\alpha = -14.00$ ev) are given in Table 5 and contrasted with those derived from ab initio calculations [27].

It is apparent from earlier calculations on NH_3 that the preferred geometry was planar ($\theta = 90°$) when $\alpha_s = \alpha_p = \alpha_H$. The effect of introducing a *s-p* separation by first-order perturbation theory is illustrated in Fig. 8. Clearly a pyramidal geometry is now preferred. Particularly influential is the greater stabilisation of the 2 a_1 (non-bonding molecular orbital).

How can a non-bonding orbital exert such a strong stereochemical influence? Since the first-order stabilisation is equal to $C_s^2\delta\alpha$, those orbitals which experience the largest changes in *s* character as θ is altered gain the greatest stabilisation. In particular for θ values of 90° and 54.73° the following changes in *s* character (C_s^2) occur:

	θ		ΔC_s^2
	90.00°	54.73°	
1 a_1	0.707^2	0.612^2	−0.125
2 a_1 (non-bonding)	0.000	0.500^2	0.250

These trends in orbital populations are clearly illustrated in Fig. 5.

In molecules such NF_3 the energy of the 2 *s* orbital of nitrogen is similar to that of the 2 *p* orbitals, which are responsible for forming the σ bond with nitrogen, but that of the 2 *p* orbitals of nitrogen is considerably higher (approx. 10 ev) [28]. In this case the same methodology (first-order perturbation theory) can be used to improve those wave functions and orbital energies associated with 2 *p* atomic orbital of nitrogen. From the zero-order wave functions of 1 a_1 and 2 a_1 developed above, the destabilisation energy from the first-order correction is:

$$[1/(\beta_s^2 + \beta_p^2)] \times (\beta_s^2/2 + \beta_p^2)$$

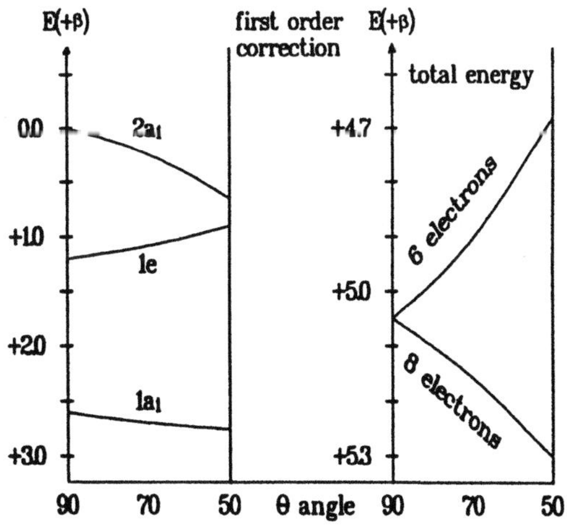

Fig. 8. The energies of molecular orbitals in MH_3 (**1**) as a function of θ angle after the first-order perturbation correction

Since the β_s is constant from our approximation, the destabilisation energy derived from raising the α_p depends on only β_p and consequently only on the cone angle θ in (**1**). The larger the magnitude of β_p, the smaller the destabilisation energy. Therefore, as θ is decreased the destabilisation energy decreases. Since the 2 *p* character in *e* orbitals is independent of θ angle, the destabilisation energies (first-order correction) are constant.

The arguments developed above show that the trends in geometric effects are the same regardless of raising α_p or lowering α_s as a perturbation.

In the transition metal complexes, the methodology can be used to evaluate the effect of $\alpha_s \neq \alpha_p \neq \alpha_d$ by lowering α_d as a perturbation. In this case, the first-order stabilisation energy is equal to $C_d^2\delta\alpha$. Therefore, the non-bonding orbital*(s)* which experience the largest changes in *d* character as the geometry is changed gain the greatest stabilisation and exert a stronger stereochemical influence.

2.7 Generation of Non-Bonding Orbitals

The discussion given in the previous sections has provided a methodology for defining the non-bonding orbitals. The Eqs. (12), (13) and (14) showed that the ligands are always located on the nodal lines of the non-bonding orbitals. The nodal properties of the non-bonding orbitals do not change even though the first-order correction from $\alpha_s \neq \alpha_p \neq \alpha_d$ is included. A more general method for defining the non-bonding orbitals can be formulated in the following equation:

$$\langle \Psi_{\text{non-bonding}} | \sigma_i \rangle = 0 \, , \tag{15}$$

where $\Psi_{\text{non-bonding}}$ is a linear combination of valence orbitals on the central atom:

$$\Psi_{\text{non-bonding}} = \sum_{l,m} a_{lm} \Psi_{lm}$$

e.g., *as* + *bp* + *cd*. Since the wave functions of the valence orbitals in the central atom are represented in terms of spherical harmonic functions of the form:

$$\Psi_{l,m} = R_{nl}(r) Y_{lm}(\theta, \varphi)$$

the $\langle \Psi_{\text{non-bonding}} | \sigma_i \rangle$ can be formulated in the following equation:

$$\langle \Psi_{\text{non-bonding}} | \sigma_i \rangle$$

$$= \sum_{l,m} a_{lm} \langle R_{nl}(r) | \sigma \rangle Y_{lm}(\theta_i, \varphi_i) \tag{16}$$

$$= \sum_{l,m} a_{lm} \langle R_{nl}(r) | \sigma \rangle [(2l + 1)/4\pi]^{1/2} C_{lm}(\theta_i, \varphi_i) \tag{17}$$

$$(\text{because } C_{lm} = [4\pi/(2l + 1)]^{1/2} Y_{lm})$$

The assumption of $S_\sigma(s) = S_\sigma(p) = S_\sigma(d)$ (see (**4**)) (S indicates the overlap integral) in fact is equivalent to that $[(2l + 1)/4\pi]^{1/2} \langle R_{nl}(r) | \sigma \rangle$ is equal for *s*, *p* and *d*. Under this assumption, the following equation is obtained:

$$\sum_{l,m} a_{lm} C_{lm}(\theta_i, \varphi_i) = 0 \qquad (18)$$

Therefore, the wave function of the non-bonding orbital can be determined by solving Eq. (18). For example, the a_1 non-bonding orbital in a square pyramid is:

$$as + bp_z + cd_{z^2}$$

Inserting the wave functions C_{lm} into above equation, we obtain:

$$a + b\cos\theta + c/2(3\cos^2\theta - 1)$$

Replacing θ by 180° and 90°, we have:

$$a - b + c = 0$$

$$a - c/2 = 0$$

Combining this with the normalisation condition, we obtain:

$$\Psi_{\text{non-bonding}} = (1/14)^{1/2} s + (9/14)^{1/2} p_z + (4/14)^{1/2} d_{z^2}$$

which is exactly the same as the one we obtained in Sect. 2.5.

In the square plannar ML_4 complex, the a_{1g} non-bonding orbital is a linear combination s and d_{z^2}, i.e.,

$$as + bd_{z^2}$$

Following the same methodology, we have:

$$a + b \times 1/2(3\cos^2\theta - 1)|_{\theta=90^\circ} = 0$$

Therefore $a - 1/2\,b = 0$. The a_{1g} non-bonding orbital which has zero overlap with the four ligands has the form of

$$(1/5)^{1/2} s + (4/5)^{1/2} d_{z^2}$$

which is also the same as the one given in Sect. 2.5.

In a discussion of hybridisation schemes [29], only the angular function Y_{lm} were used in Eq. (18) instead of C_{lm} to determine the non-bonding orbitals. That corresponds to the approximation that $\langle R_{nl}(r)|\sigma\rangle$ is the same for s, p and d. It can be correlated to the molecular orbital method by assuming $S_\sigma(s) : S_\sigma(p) : S_\sigma(d) = 1 : (3)^{1/2} : (5)^{1/2}$. It is apparent that the Y_{lm} approximation tends to overemphasise the overlap of d orbitals with ligands. Nevertheless, the relative sign of the orbital mixings in the non-bonding orbitals are reproduced as are the trends associated with the variation of the nodal characteristics as the d, s and p contributions are varied. For a spherical co-ordination compound, the effect of the above assumptions on the solutions of the non-bonding orbitals is limited since the non-bonding orbitals do not involve much admixture between s, p and d.

The other method to generate a non-bonding orbital is based on the spherical and complementary nature of the electron density in co-ordination compounds and the localisation procedure described previously [11]. The linear combination of the octahedral symmetry adapted ligand linear combinations $(1/6)S^{\sigma} + (1/2)P_0^{\sigma} + (1/3)D_0^{\sigma}$ is localised exclusively at position 1 of the octahedron (see Fig. 1) and the remaining five linear combinations are delocalised over the other five locations and may be constructed to be equivalent to the symmetry adapted linear combinations for a square pyramid. The localisation procedure was developed by Mingos and Hawes [11]. The linear combination $as + bp_z + cd_{z^2}$ produces a region of electron density in the position 1 of the octahedron. When the coefficients a, b and c are equal to those $(1/6)S^{\sigma} + (1/2)P_0^{\sigma} + (1/3)D_0^{\sigma}$ gives $a = (1/6)^{1/2}$, $b = (1/2)^{1/2}$ and $c = (1/3)^{1/2}$. The orbital $(1/6)^{1/2}s + (1/2)^{1/2} + (1/3)^{1/2}d_{z^2}$ maximises the electron density in the position 1. An electron pair occupying this orbital localised on the central atom will be essentially non-bonding with respect to the bonding molecular orbitals associated with the square pyramidal co-ordination polyhedron.

2.8 Summary

The arguments developed above provide some fundamental principles governing the nature of the molecular orbitals of ML_n compounds. Although the **m − n rule** and the **pairing theorem** are based on the assumption of $\alpha_s = \alpha_p = \alpha_d$ (and $= \alpha_l$ for the latter), the effects of unequal Coulomb integrals ($\alpha_s \neq \alpha_p \neq \alpha_d$) can be treated using a perturbation correction and most conclusions derived from them retain their generality. The following general points can be made from the discussions of the previous sections.

(1) The number of non-bonding orbitals in ML_n complexes is equal to $|m - n|$ if the number of valence orbitals in the central atom is m.

(2) The group theoretical aspects for defining the number of non-bonding orbitals can be applied to any ML_n compound.

(3) The ligands are always located on the nodal lines, planes or cones of the non-bonding orbitals. The nodal properties do not change even when the first-order correction arising from $\alpha_s \neq \alpha_p \neq \alpha_d$ is included.

(4) The non-bonding orbitals can be classified into following categories:

(a) pure atomic orbitals when the ligands are all located on their nodal planes or cones e.g., $BH_3(p_z)$ and $ML_6(d_{xz}, d_{yz}, d_{xy})$. Further examples of these types of non-bonding orbitals are tabulated in Table 6.

(b) hybrids (mixing of *spd, sp* or *dp*) whose nodal lines or planes coincide with the metal-ligand bond directions. In molecules belonging to C_{nv} point groups, where s, p_z and d_{z^2} belong to the a_1 irreducible representation, a hybrid $s + \lambda p_z + \lambda'' d_{z^2}$ can be defined which has nodes in the ligand directions. Such a hybrid points towards the vacant co-ordination site of the polyhedron. In main group chemistry the hybrid is generally occupied by an electron pair, whereas in transition metal compounds it is generally vacant. This difference can be traced back to the relative energies of the valence orbitals i.e., $|\alpha_s| > |\alpha_p| \gg |\alpha_d|$ in the former and $|\alpha_d| > |\alpha_s| \gg |\alpha_p|$ in the latter. All main group molecules that have a stereochemically acitve non-bonding orbital have geometries belonging to the C_{nv} point groups. Specific examples include $NH_3(C_{3v})$, $SF_4(C_{2v})$, $PbO(C_{4v})$ and $BrF_5(C_{4v})$. For higher co-ordination molecules ($n > 3$) the adoption of these C_{nv} geometries in preference to higher symmetries

Table 6. Non-bonding orbitals localised on the central atom and not hybridised

Geometry	*p* orbitals	*d* orbitals
Linear ML_2	p_x, p_y	$d_{x^2-y^2}, d_{xy}, d_{xz}, d_{yz}$
Trigonal ML_3	p_z	d_{xz}, d_{yz}
Tetrahedral ML_4		$d_{x^2-y^2}, d_{z^2}$
Square planar ML_4	p_z	d_{xy}, d_{xz}, d_{yz}
Trigonal bipyramidal ML_5		d_{xz}, d_{yz}
Octahedral ML_6		d_{xy}, d_{xz}, d_{yz}
Pentagonal bipyramidal ML_7		d_{xz}, d_{yz}
Dodecahedral ML_8		$d_{x^2-y^2}$

(more spherical geometries) also avoids the necessity of occupying a metal-ligand antibonding molecular orbital. For example, in tetrahedral SH_4 molecule the additional electron pair would occupy a M − L σ antibonding molecular orbital which if $|\alpha_s| \gg |\alpha_p|$ is of a_1 symmetry, and an out-of-phase combination of s and S^σ. Consequently, the $T_d \rightarrow C_{2v}$ geometric distortion has the effect of converting an antibonding interaction into a non-bonding interaction. This is an important and ubiquitous driving force for geometric distortions not only in simple molecules but also in infinite solids [30]. However, the driving force is only a significant one if the highest occupied molecular orbital is strongly antibonding. If it is not then the greater stabilisation energy associated with the transformation of an antibonding into a non-bonding orbital does not outweigh the greater destabilising ligand-ligand interactions which result from the lower symmetry structure.

As the co-ordination number increases, the antibonding character of the $s - S^\sigma$ molecular orbital becomes smaller and there are many examples of molecules with 6, 8 and 12 co-ordination which do not exhibit the stereochemical activity of the lone pair. Specific examples include $[SbX_6]^{3-}$ (X = Cl, Br or I); $[MX_6]^{2-}$ (M = Se or Te; X = Cl, Br or I) which have seven electron pairs but remain octahedral and $[XeF_8]^{2-}$ which has nine electron pairs and is square-antiprismatic. As Ng and Zuckerman [30] have noted, the absence of a stereochemical effect with the lone-pair orbital is particularly prevalent for infinite structures with high co-ordination numbers and octahedral, cubic or dodecahedral geometries. In such situations although the additional electron pair is occupying an $s - S^\sigma$ antibonding orbital, it is unlikely to be strongly antibonding if $|\alpha_s| \gg |\alpha_p|$ and the β_s resonance integral is small because of the core-like nature of the s orbital.

In the dihedral point groups s and p_z no longer have the same symmetry transformation properties and $s \pm \lambda d_{z^2}$ non-bonding hybrids only can be generated. In the linear molecules (see Table 7) the hybrids concentrate density in the direction perpendicular to the metal-ligand bond directions. In the trigonal prism and the square antiprism the ligands lie near the cone of the hybrid. These hybrids which are obtained from Eq. (18) are illustrated in Fig. 9.

In finite dihedral point groups $d_{x^2-y^2}$ and d_{xy} transform in the same manner as p_x and p_y and, for example, in trigonal planar and trigonal bipyramidal molecules hybrids of the type illustrated in Fig. 10 are generated.

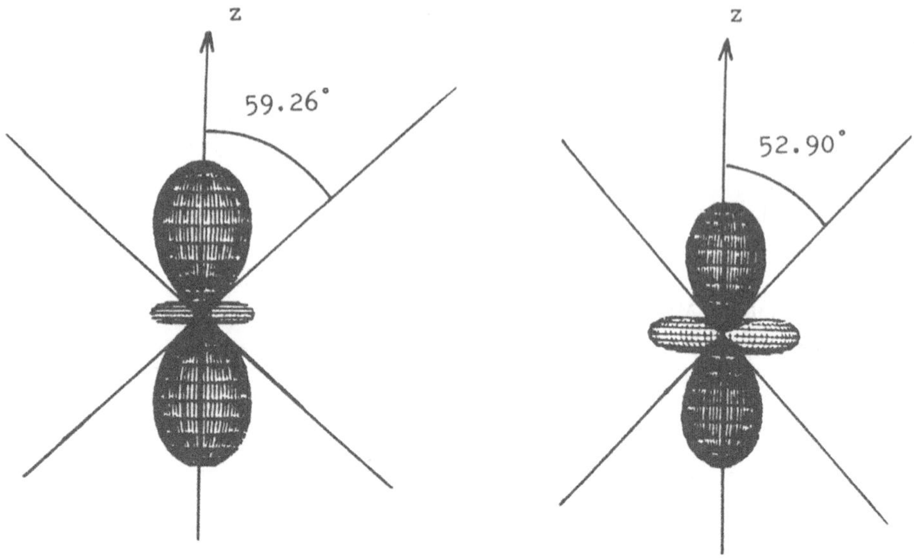

$0.107s+0.994d_{z^2}$

(a)

$0.046s-0.999d_{z^2}$

(b)

Fig. 9. The a_1 non-bonding orbital for a square antiprismatic transition metal ML_8 complex

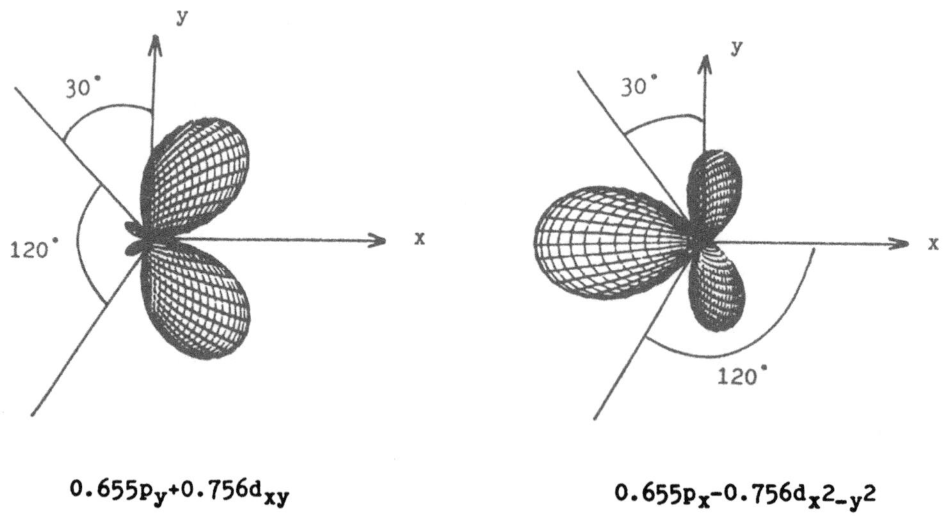

$0.655p_y+0.756d_{xy}$

$0.655p_x-0.756d_{x^2-y^2}$

Fig. 10. The *e* pair of $d-p$ hybrid non-bonding orbitals for the trigonal planar and the trigonal bipyramidal molecules

Table 7. Non-bonding hybrids in dihedral point groups

	$s \pm \lambda d_{z^2}$	$d_{x^2-y^2} \pm \lambda p_x$	$d_{xy} \pm \lambda p_z$
Linear ML_2	yes		
Trigonal planar ML_3	yes	yes	yes
Square planar ML_4	yes		
Trigonal bipyramidal ML_5		yes	yes
Trigonal prismatic ML_6	yes	yes	yes
Square antiprismatic ML_8	yes		

In the tetrahedral point group (d_{xz}, d_{yz}, d_{xy}) and (p_x, p_y, p_z) both transform according to the t_2 irreducible representation and generate three *d-p* hybrids which point towards the vertices of a cube not utilized by the tetrahedron (see Fig. 11).

(c) The non-bonding orbitals are localised on the ligands when the number of ligand orbitals is larger than that of valence orbitals on the central atom e.g., PF_5 and SF_6. This localisation limits the occurrence of such non-bonding orbitals to molecules with very electronegative ligands, e.g., F and O.

(d) π-localised non-bonding orbitals on the ligands can also occur, e.g., MnO_4^{2-} (T_d point group) since the 12 *p* orbitals in oxygen transform as $a_1 + 2\, t_2 + e + t_1$ and the valence orbitals on the central atom as $a_1 + 2\, t_2 + e$, t_1 are the π-localised non-bonding orbitals.

(5) The comparison between Figs. 4 and 8 indicates that a filled non-bonding orbital*(s)* does exert a very strong stereochemical influence, when $|\alpha_s| > |\alpha_p|$. This stereochemical activity is associated with the redistribution of *s* and *p* electron density which accompanies the geometric distortion. Although the discussion above has focussed on the occurrence of non-bonding orbitals in molecular compounds, it should be noted that the non-bonding character has been defined with respect to the σ-radial orbitals. Such non-bonding orbitals which have nodes in the metal-ligand directions are ideally suited for forming π-bonds with orbitals of the appropriate symmetry. These types of interactions are maximised when the non-bonding orbital has a minimal contribution from the *s* orbital. For example, in planar MX_3 the maximum π-bonding is achieved for the planar molecule which has a non-bonding p_z orbital on M rather than the pyramidal molecule where the non-bonding orbital is a s-p_z hybrid. The superior p_z-X_π ligand overlap integrals for the planar geometry reinforce this preference.

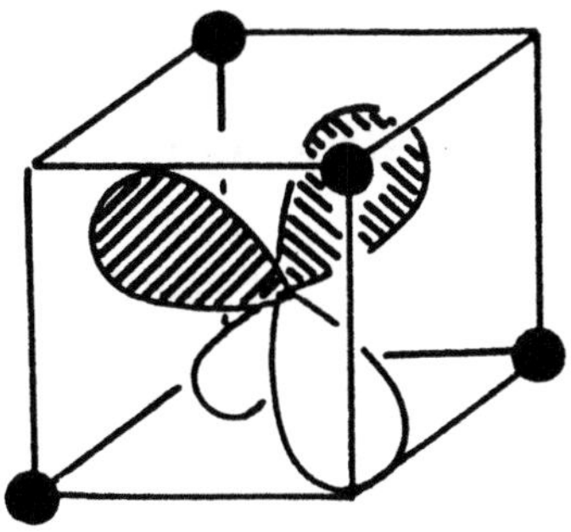

Fig. 11. $p_z - d_{xy}$ hybrid orbital in the tetrahedral point group

In transition metal complexes similar arguments apply to d-s hybrids, but d_{xz}-p_x hybrids can result in very favorable π-bonding effects between the metal and the ligand.

In molecules where m − n = 2 there are, by the "n − m rule", two non-bonding orbitals. These can be a degenerate e set resulting from the occurrence of an odd number of e representations in the point group. Alternatively, the two non-bonding orbitals must belong to different irreducible representations. If they belonged to the same they would interact strongly especially since they have similar energies to produce a bonding and an antibonding orbital. In main group molecules with two non-bonding electron pairs the first is accommodated in an orbital of a_1 ($s + \lambda p_z + \lambda' d_{z^2}$) symmetry and the second in a p_y orbital orthogonal to the ligand plane. For example, OH_2, ClF_3 and XeF_4 all have planar geometries which satisfy this condition, although in the latter example the first electron pairs occupies $s + \lambda d_{z^2}$ rather $s + \lambda p_z + \lambda' d_{z^2}$.

(6) Effects associated with the radial parts of the wave functions. The discussion presented above has emphasised the importance of the nodal characteristics of the molecular orbitals and many of the conclusions have depended on the assumption that $\beta_s \approx \beta_p \approx \beta_d$. This is a good approximation when the radial parts of the wave functions lead to comparable overlaps with the ligand orbitals, but for many real molecules this approximation begins to break down. For example, for a first-row main group atom $\beta_s \approx \beta_p$ but for heavier atoms $|\beta_s| < |\beta_p|$ and this has important consequences for the geometries of molecules with stereochemically active lone pair. In particular there is more s character in the non-bonding orbital and the geometries show greater distortions. Furthermore, for molecules such as the inert gas molecules, XeF_2, XeF_4 and XeF_6, the s orbitals on the central atom are effectively core-like because of their high ionisation potential i.e., $|\beta_s| \ll |\beta_p|$.

For transition metal atoms the nd, (n + 1)s and (n + 1)p valence orbitals have very different radial distribution functions. In particular, the nd orbitals are much more contracted than (n + 1)s and (n + 1)p. Therefore, the nd orbitals do not overlap strongly with the ligand orbitals unless the metal-ligand distances are short. This is achieved either with small ligands e.g., H, or ligands which form multiple bonds with the metal e.g., CO, O or N. When the ligands do not overlap very strongly then the d orbitals approximate to non-bonding. Chemically this is reflected in the occurrence of high-spin and low-spin complexes and variable occupations of the d-manifold orbitals. In lanthanide complexes the 4f orbitals are also very contracted and almost core-like. Consequently, their overlap with ligands is small and they exert no stereochemical influence.

3 Non-Bonding Orbitals in Hydrocarbons

3.1 *Alternant Hückel Hydrocarbons*

A conjugated organic system is described as *alternant* if the atoms in it can be divided into two groups *starred* and *unstarred* in such a way that no two atoms are of like parity i.e., which belong to the same group, are directly linked. Some typical examples of *even* and

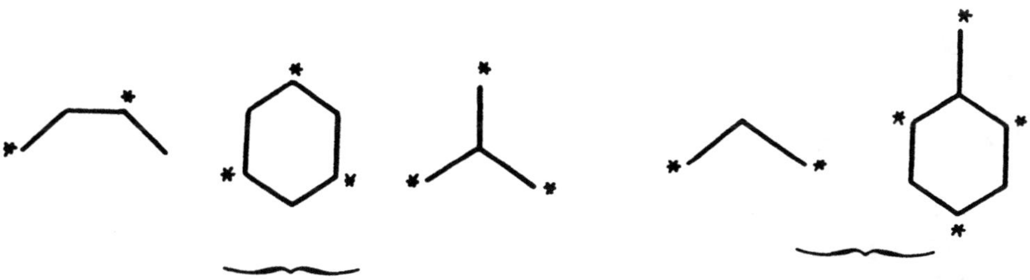

Fig. 12. Some typical examples of even and odd alternant hydrocarbons

odd alternant hydrocarbons are illustrated in Fig. 12. When the number of atoms in the two sets differ, the more numerous set is called the starred set. Within the Hückel approximation such alternant systems having the following important properties which collectively are described as the Coulson-Rushbrooke Pairing Theorem [19].

(1) The molecular orbitals of an even AH (AH, alternant hydrocarbon) occur in pairs with energies $\alpha \pm E_\mu$, α being the Coulomb integral for carbon [15].

(2) If one of the MO's Φ_μ^+ of a pair (energy $\alpha + E_\mu$) is given by:

$$\Phi_\mu^+ = \sum_i^* a_{\mu i}^* \phi_i^* + \sum_j^0 a_{\mu j}^0 \phi_j^0 ,$$

where ϕ_i^* is the *p* atomic orbital of a starred atom and ϕ_j^0 that of an unstarred atom then the twin MO Φ_μ^- (energy $\alpha - E_\mu$) is given by:

$$\Phi_\mu^- = \sum_i^* a_{\mu i}^* \phi_i^* - \sum_j^0 a_{\mu j}^0 \phi_j^0$$

Therefore, the coefficients of the paired molecular orbitals are numerically the same, but the signs of one set of MO's are reversed in going from one MO to the other.

(3) In an odd AH where the total number of MO's is correspondingly odd, all but one of the MO's is paired as in (1) and (2) above. The odd MO has energy α and is confined to the starred atoms.

$$\Phi_0 = \sum_i^* a_{0i}^* \phi_i^*$$

These conclusions can also be interpreted in terms of the "m − n rule" developed above and are a direct consequence of the properties of the Eq. (1) matrix. The Pairing Theorem does not completely define the spectrum of molecular orbitals in an alternant hydrocarbon, because (1) is consistent with the presence of an even number of molecular orbitals. Whilst this might generally mean that the number of non-bonding molecular orbitals is 0, there are instances when there are 2, 4, etc. For example, both butadiene and cyclo-butadiene are even alternant hydrocarbons, but the former has no non-bonding molecular orbitals and the latter 2 (see Fig. 13).

Group theory can be used to resolve this distinction, because the sets of paired molecular orbitals defined above are related by specific symmetry relationships. If Φ_μ^+

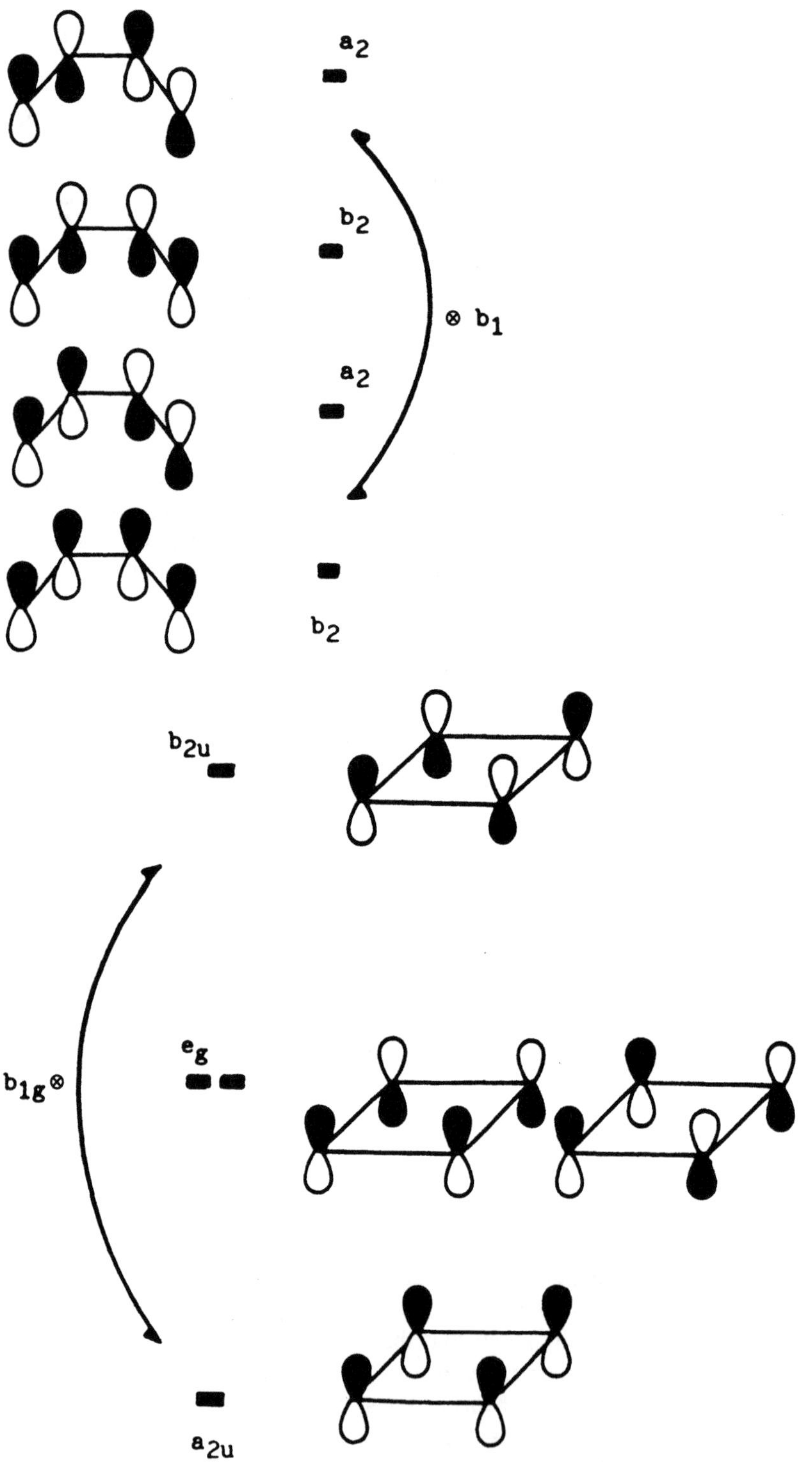

Fig. 13. Molecular orbitals for *cis*-butadiene and cyclo-butadiene

belongs to the irreducible representation Γ_+, and Φ_μ^- belongs to the irreducible representation Γ_-, they are related by the following direct products:

$$\Gamma_+ = \Gamma_A \otimes \Gamma_-$$

$$\Gamma_- = \Gamma_A \otimes \Gamma_+$$

where Γ_A is a one-dimensional representation which interconverts the starred and unstarred atomic positions. It has the character of +1 for those operations which merely interconvert the positions of starred atoms and −1 for those operations which interconvert starred and unstarred positions.

For example, for *cis*-butadiene (C_{2v} point group) Γ_A corresponds to the b_1 representation (the carbon atoms lie in the xz plane) and the paired molecular orbitals shown in Fig. 13 are related in the following manner:

$$\Gamma_- = \Gamma_A \otimes \Gamma_+$$

$$\Gamma_- = b_1 \otimes b_2 = a_2$$

$$\Gamma_- = b_1 \otimes a_2 = b_2$$

Similarly for cyclo-butadiene (D_{4h} point group) Γ_A transforms according to the b_{1g} representation and the a_{2u} and b_{2u} bonding and antibonding molecular orbitals shown in Fig. 13 are related by the following direct product:

$$b_{2u} = b_{1g} \otimes a_{2u}$$

For reasons which will be discussed below it is highly significant that the non-bonding molecular orbitals remain unaffected by the multiplication.

$$e_g = b_{1g} \otimes e_g$$

The Γ_A irreducible representation is not a characteristic of the point group, but reflects the symmetry transformation properties of the starred and unstarred atoms. For example, *cis*-butadiene and allyl, C_3H_5, both belong to the C_{2v} point group but $\Gamma_A = b_1$ for the former and $\Gamma_A = a_1$ for the latter. Indeed, in any molecule where there is an atom on the principal symmetry axis the Γ_A will correspond to the totally symmetric representation and the paired molecular orbitals will have the same symmetry transformation properties i.e.,

$$\Gamma_- = a_1 \otimes \Gamma_+ = \Gamma_+ .$$

The examples described above can be generalised into three distinct situations. We define the representations as conjugate when they are transformed by the representation, Γ_A, corresponding to the alternant operator, $\hat{A}$.

(a) Γ_- is always distinct from its conjugate Γ_+. *Cis*-butadiene provided an example of this behaviour.

(b) Γ_- is always self-conjugate with Γ_+. This occurs when Γ_A belongs to the totally symmetric representation e.g., allyl.

(c) Some Γ_- are self-conjugate and some are not e.g., cyclo-butadiene, where b_{2u} and a_{2u} are conjugate and e_g self-conjugate.

These symmetry considerations become highly significant when combined with the pairing theorem, because if *the total number of irreducible representations spanned by the atomic basis functions contains an odd number of self-conjugate representations it has a set of non-bonding molecular orbitals equal to the sum of the degeneracies of the self-conjugate representations.*

For example, cyclo-butadiene has two non-bonding molecular orbitals arising from the odd number of self-conjugate e_g representations and allyl has one non-bonding molecular orbital of a_2 symmetry deriving from the odd number (1) of a_2 irreducible representations.

In benzene, the alternant operator, Â, corresponds to the b_{1u} representation, and the paired molecular orbitals shown in Fig. 14 are related by the following direct products:

$$\Gamma_- = \Gamma_A \otimes \Gamma_+$$

$$b_{2g} = b_{1u} \otimes a_{2u}$$

$$e_{2u} = b_{1u} \otimes e_g$$

There are no self-conjugate molecular orbitals and therefore no non-bonding molecular orbitals. Without labouring the point it is clear that the pairing theorem together with alternant symmetry operator relationships define the 4 n + 2 Hückel rule for cyclic delocalisation.

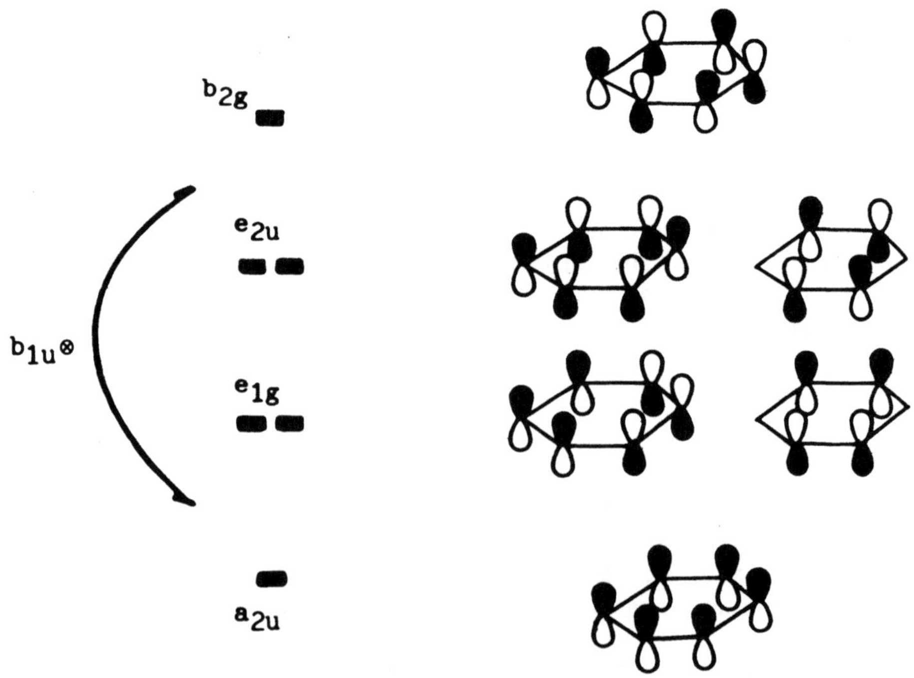

Fig. 14. Molecular orbitals for benzene

3.2 *Alternant Möbius Hydrocarbons*

Alternant Möbius hydrocarbons conform to the pairing theorem even though the topology of the orbital interactions is different. The molecular orbitals of a Möbius ring of π-orbitals are analysed using the double group $C_n \otimes R$, because the n successive 2 π/n and twists of π/n of the p_π basis functions result in $\phi_1 \rightarrow -\phi_1$ after a rotation of 2 π, and $\phi_1 \rightarrow \phi_1$ after a total rotation of 4 π(720°). For an even Möbius system the p_π functions generate E′, E″, E‴, . . ., representations, where the total number of E representations is equal to n/2. For example, for Möbius cyclo-butadiene E′ and E″ and Möbius benzene E′, E″, E‴. The alternant operator Â transforms as B in this pseudo-cyclic group and leads to the following relationships:

Möbius cyclo-butadiene	*Möbius* cyclo-benzene
$E'' = B \otimes E'$	$E''' = B \otimes E'$
	$E'' = B \otimes E''$
no self-conjugate representations	E″ self-conjugate representation

Consequently, the requirement that non-bonding orbitals be either completely filled or empty leads to the Möbius 4 n rule for cyclic delocalisation.

The combination of Hückel and Möbius generalisations can be utilised to restate the Woodward-Hoffmann rules [31], in a manner that has been popularised by Zimmermann [32].

4 Non-Bonding Orbitals in Cluster Compounds

Three-dimensional polyhedral molecules such as the borane anions, $B_nH_n^{2-}$, are characterised by a total of n + 1 skeletal electron pairs or 4 n + 2 valence electrons in total [33]. The isolobal analogy between the BH and ML_3 fragments has provided a theoretical relationship between the borane and the transition metal cluster compounds [34]. A detailed analysis of the molecular orbitals of deltahedral transition metal clusters showed that they also have a total of n + 1 bonding skeletal molecular orbitals, which is the same as that in the borane clusters, and 6 n orbitals that are either non-bonding or involved in the metal-ligand bonding. Therefore, such clusters are characterised by 14 n + 2 valence electrons in total.

It is apparent that deviations from the above electron counts in cluster compounds result from an occurrence of skeletal non-bonding orbitals. The skeletal non-bonding orbitals arise either from topological factors associated with deltahedral geometries or the presence of missing vertices in *nido-* and *arachno*-structures.

4.1 Expansion of the Cluster Skeletal Molecular Orbitals in Terms of the Tensor Surface Harmonic Functions

Recently, an elegant approach to the bonding in clusters has been developed by Stone [35], whose **Tensor Surface Harmonic Theory** derives cluster skeletal molecular orbitals as expansions of vector surface harmonic functions. The skeletal molecular orbitals are generated from basis sets with 0(σ) and 1(π) nodes with respect to a radial vector passing through the atoms.

For a basis set of σ orbitals, the LCAO expansions take a form that is exactly the same as the expansion for ligand σ-orbitals, i.e.,

$$\Psi^{\sigma}_{LM} = L^{\sigma}_{M} = N \sum_{i} C_{LM}(\theta_i, \varphi_i)\sigma_i \tag{19}$$

For a basis set of π-orbitals, the LCAO expansions are derived from the sets of π_i^θ and π_i^φ orbitals which are defined in Fig. 15 and have the following form:

$$\Psi^{\pi}_{LM} = L^{\pi}_{M} = N \sum_{i} [V^{\theta}_{LM}(\theta_i, \varphi_i)\pi_i^{\theta} + V^{\varphi}_{LM}(\theta_i, \varphi_i)\pi_i^{\varphi}] \tag{20}$$

where $(V^{\theta}_{LM}, V^{\varphi}_{LM})$ are Vector Surface Harmonics. For M > 0 the real forms of the Vector Surface Harmonics $V^{\theta,\varphi}_{LMs}/V^{\theta,\varphi}_{LMc}$ are used and listed in Table 8. The above L^π wave functions are called even parity functions. The corresponding odd parity wave functions $\bar{\Psi}^{\pi}_{LM}(\bar{L}^{\pi})$ are related to $\Psi^{\pi}_{LM}(L^{\pi})$ by a 90° anticlockwise rotation of the tangential vectors $(\pi_i^\theta, \pi_i^\varphi)$ about the radial vector, at each vertex. Therefore, the wave functions of $\bar{L}^\pi$ have the following form:

$$\bar{\Psi}^{\pi}_{LM} = \bar{L}^{\pi}_{M} = N \sum_{i} [V^{\varphi}_{LM}(\theta_i, \varphi_i)\pi_i^{\theta} - V^{\theta}_{LM}(\theta_i, \varphi_i)\pi_i^{\varphi}] \tag{21}$$

Stone [35] showed that for deltahedral clusters the L^π MOs are generally bonding. The $\bar{L}^\pi$ MOs are, therefore, generally antibonding, since the parity inversion operation leads to a reversal of the bonding properties of L^π and $\bar{L}^\pi$. He used the methodology of Tensor Surface Harmonic theory to derive the n + 1 skeletal electron pair rule for

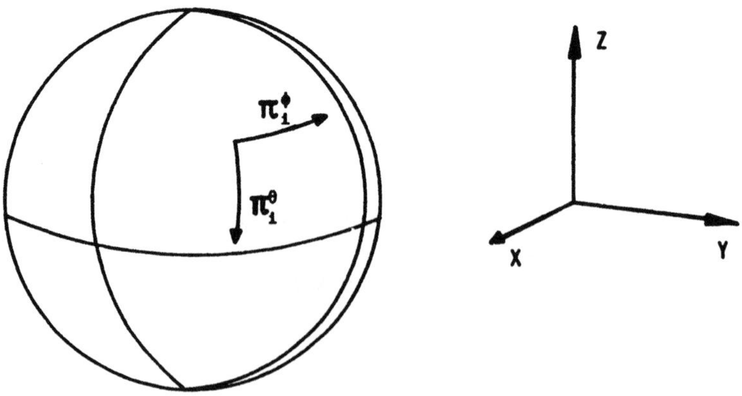

Fig. 15. Unit vectors representing the *p* orbitals on the surface of the cluster sphere

Table 8. Even vector harmonics

Y^{π}_{lm}	V^{θ} component	V^{φ} component
p^{π}_{0}	$-\sin\theta$	0
p^{π}_{1s}	$\cos\theta\sin\varphi$	$\cos\theta$
p^{π}_{1c}	$\cos\theta\cos\varphi$	$-\sin\varphi$
d^{π}_{0}	$-3/2\sin 2\theta$	0
d^{π}_{1s}	$3^{1/2}\cos 2\theta\sin\varphi$	$3^{1/2}\cos\theta\cos\varphi$
d^{π}_{1c}	$3^{1/2}\cos 2\theta\cos\varphi$	$-3^{1/2}\cos\theta\sin\varphi$
d^{π}_{2s}	$(3/4)^{1/2}\sin 2\theta\sin 2\varphi$	$3^{1/2}\sin\theta\cos 2\varphi$
d^{π}_{2c}	$(3/4)^{1/2}\sin 2\theta\cos 2\varphi$	$-3^{1/2}\sin\theta\sin 2\varphi$
f^{π}_{0}	$-3/2\sin\theta(5\cos^2\theta-1)$	0
f^{π}_{1s}	$(3/8)^{1/2}\cos\theta(15\cos^2\theta-11)\sin\varphi$	$(3/8)^{1/2}(5\cos^2\theta-1)\cos\varphi$
f^{π}_{1c}	$(3/8)^{1/2}\cos\theta(15\cos^2\theta-11)\cos\varphi$	$-(3/8)^{1/2}(5\cos^2\theta-1)\sin\varphi$
f^{π}_{2s}	$(15/4)^{1/2}\sin\theta(3\cos^2\theta-1)\sin 2\varphi$	$(15/4)^{1/2}\sin 2\theta\cos 2\varphi$
f^{π}_{2c}	$(15/4)^{1/2}\sin\theta(3\cos^2\theta-1)\cos 2\varphi$	$-(15/4)^{1/2}\sin 2\theta\sin 2\varphi$
f^{π}_{3s}	$(3/4)(10)^{1/2}\sin^2\theta\cos\theta\sin 3\varphi$	$(3/4)(10)^{1/2}\sin^2\theta\cos 3\varphi$
f^{π}_{3c}	$-(3/4)(10)^{1/2}\sin^2\theta\cos\theta\cos 3\varphi$	$(3/4)(10)^{1/2}\sin^2\theta\sin 3\varphi$

deltahedral boranes. Each of the n BH fragments possesses one σ-type *sp* hybrid, yielding nL^{σ} MOs in total, one of which (S^{σ}) is strongly bonding because it possesses no angular nodes. In addition, each fragment has two π-type *p* orbitals which generate nL^{π} bonding and $n\bar{L}^{\pi}$ antibonding cluster orbitals. Mixing between L^{σ} and L^{π} orbitals with the same L values results in one bonding combination and one antibonding combination for every matched pair of L^{σ}/L^{π} orbitals. In total, therefore, there are n + 1 cluster bonding MOs (S^{σ} and nL^{σ}/L^{π}) since there is no S^{π} orbital.

4.2 Pairing Principle and Group Theoretical Aspects

Stone [35] has shown how every bonding overlap is converted to an antibonding overlap by the parity operation, and vice versa. Thus in the absence of σ − π interaction, there would be a mirror relationship between the energies of the L^{π} and $\bar{L}^{\pi}$ orbitals. This is the pairing principle in TSH theory. Even after σ − π interactions are allowed for it has been shown that an approximate pairing persists in the frontier orbital region [36]. As with the discussion above for coordination compounds, this pairing principle is particularly useful for defining the non-bonding orbitals in cluster compounds especially when combined with the group theoretical aspects of the problem.

In terms of irreducible representations, the symmetry of an L^{π} set is converted to that of the $\bar{L}^{\pi}$ set by multiplication by the odd-parity pseudo-scalar representation Γ°_{u} [37].

$$\Gamma(\bar{L}^{\pi}) = \Gamma(L^{\pi}) \otimes \Gamma^{\circ}_{u}$$

Γ°_{u} is a one-dimensional representation with characters +1 under proper rotations and −1 under improper rotations.

Fowler [36] has elegantly demonstrated that for the point groups C_i, C_s, C_{2v}, $D_{(2n+1)}d$, C_{nh}, $S_{(4n-2)}$, D_{nh}, T_h, O_h, I_h and R_{3h} $\Gamma(L^\pi)$ is always distinct from its conjugate $\Gamma(\bar{L}^\pi)$. In contrast, in the point groups C_n, D_n, T, O and I, $\Gamma(L^\pi)$ and $\Gamma(\bar{L}^\pi)$ are always self-conjugate.

Therefore, in the vast majority of deltahedral clusters which belong to these point groups, the total representation $\Gamma_{\pi+\bar{\pi}}$ splits into two distinct sub-shells and there are n bonding and n antibonding molecular orbitals. Since such boranes also have a single radial bonding molecular orbital belonging to the totally symmetric representation, they are characterised by a total of n + 1 skeletal electron pairs or 4 n + 2 valence electrons in total.

However, in the point groups $C_{nv}(n > 2)$, S_{4n}, D_{2nd} and T_d only some of the L^π combinations are self-conjugate. Using reasoning similar to that developed above for co-ordination compounds, it can be demonstrated that if $\Gamma_{\pi+\bar{\pi}}$ contains an odd number of self-conjugate e representations then the deltahedron will have a non-bonding *e* set and the polyhedron will be associated with either n or n + 2 skeletal electron pairs.

4.3 Non-Bonding Orbitals in Closo Deltahedral and Three-Connected Cluster Compounds

As indicated above, the irreducible representations spanned by the L^π and $\bar{L}^\pi$ tangential molecular orbitals in *closo* deltahedral clusters may contain an odd number of self-conjugate representations. For example, for the tetrahedron Γ_u^0 transforms as a_2, and the $\Gamma_{\pi+\bar{\pi}}$ tangential molecular orbitals span the representations $t_1 + e + t_2$. Since

$$\Gamma(\bar{L}^\pi) = \Gamma_u^0 \otimes \Gamma(L^\pi)$$

$$t_1 = a_2 \otimes t_2$$

$$e = a_2 \otimes e$$

there is an odd number of self-conjugate *e* representations and the tetrahedron has a non-bonding *e* set.

The relevant *e* set of non-bonding orbitals for the tetrahedron is illustrated in Fig. 16. The two components of the *e* set are related by the parity operation, $\hat{P}$, i.e., a local

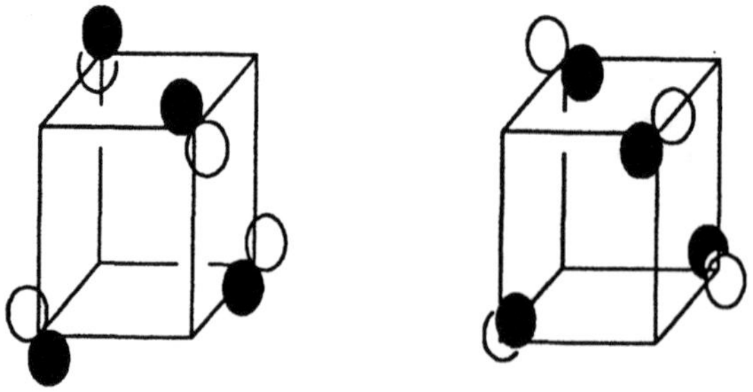

parity related e pair

Fig. 16. The *e* non-bonding molecular orbitals in the tetrahedral cluster rotation of 90° in the same sense and are therefore self-conjugate. Since they can be equally well defined as D^{π} or $\bar{D}^{\pi}$ molecular orbitals, they must be non-bonding if $\beta^{\sigma} = \beta^{\pi}$.

A general group theoretical criterion, based on the pairing principle in TSH theory, has been presented by Fowler [36] and Johnston and Mingos [38]. Fowler concluded that an odd number of self-conjugate *e* representations in *closo* deltahedral clusters arise in two cases:

(1) T or T_d clusters with an odd number of sets of four equivalent cage atoms;

(2) C_m or C_{mv} ($m \geq 3$) clusters with an odd number of cage atoms on the C_m-axis.

Subsequently, Johnston and Mingos [38] pointed out that it is not geometrically feasible to have a *closo*-deltahedron with an odd number of cage atoms on the C_m axis with $m > 3$, because square, pentagonal and higher faces would be generated. They found that for deltahedra, the occurrence of an odd number of self-conjugate representations occurs as a consequence of there being one atom on each C_3 rotation axis of a polyhedron belonging to one of the point groups C_3, C_{3v}, T or T_d. These clusters are defined as polar deltahedra with $3p + 1$ cluster vertices. In such clusters there are *p* triangles of symmetry related atoms lying in planes perpendicular to the C_3 axis and one polar atom lying on the C_3 axis. For each of the *p* triangles of the symmetry related atoms the linear combinations of $(\pi^{\theta}, \pi^{\varphi})$ consist of two *e* irreducible representations. The total number of *e* irreducible representations spanned by L^{π} and $\bar{L}^{\pi}$ are $2p$ plus one *e* pair from the polar atom. Therefore they are characterised by n skeletal bonding molecular orbitals and one e set of non-bonding orbitals and have n or n + 2 skeletal electron pairs. Examples illustrating the above conclusion are the tetrahedral clusters ($p = 1$: B_4Cl_4 with n skeletal electron pairs; C_4R_4 with n + 2 pairs), capped octahedron ($p = 2$: $Os_7(CO)_{21}$ [39] with n skeletal electron pairs) and the 10-atom metalloborane $[H(PPh_3)(Ph_2PC_6H_4IrB_9H_8)]$ ($p = 3$: with n skeletal electron pairs) [40].

The three-connected cluster compounds are characterised by a total of 3n/2 skeletal electron pairs which correspond to the localisation of each electron pair on an edge of the polyhedron. Johnston and Mingos [41] have shown that in these polyhedra there are (n/2) + 2 strongly skeletal bonding molecular orbitals and n − 2 approximately non-bonding skeletal molecular orbitals. The latter correspond to a weakly bonding set of $(n/2 - 1)L^{\pi}$ MOs, and by the pairing principle an equal number of weakly antibonding $\bar{L}^{\pi}$ MOs. These non-bonding orbitals are not a consequence of an odd number of *e* representations since most of the three-connected clusters belong to the D_{nh} point groups. In all the prismatic three-connected polyhedra the irreducible representations spanned by the L^{π} and $\bar{L}^{\pi}$ tangential molecular orbitals are always distinguished from each other and are not self-conjugate. In the prismatic structures, Johnston and Mingos noted that the non-bonding nature of these orbitals arise as a consequence of being antibonding within the upper and lower layers of atoms and bonding across the equator or vice versa. They found that $D^{\pi}_{\pm1}(e'')$ orbitals are non-bonding for a trigonal prism, $D^{\pi}_{\pm1}(e'')$ and D^{π}_{2c} orbitals for a cube, $D^{\pi}_{\pm1}(e''_1)$ and $F^{\pi}_{\pm2}(e''_2)$ for a pentagonal prism and so on. More generally, for a prismatic cluster with n = 2 k atoms we conclude that when k = odd $D^{\pi}_{\pm1}$, $F^{\pi}_{\pm2}$, $G^{\pi}_{\pm3}$, ..., $\Lambda^{\pi}_{\pm(\Lambda-1)}$ ($\Lambda = L_{max}$) and their counterparts $\bar{L}^{\pi}$ make a total of n − 2 approximately non-bonding orbitals. When k = even $D^{\pi}_{\pm1}$, $F^{\pi}_{\pm2}$, $G^{\pi}_{\pm3}$, ..., $\Lambda^{\pi}_{\pm(\Lambda-1)}$, $\Lambda^{\pi}_{\Lambda c}$ ($\Lambda = L_{max}$) and their counterparts $\bar{L}^{\pi}$ again make a total of n − 2 approximately non-bonding orbitals.

4.4 Non-Bonding Orbitals in Nido Structures

The TSH theory is based on a free electron model derived from the solution of the Schrödinger equation for a particle on a sphere. The grossly non-spherical nature of *nido* structures means that their skeletal molecular orbitals are no longer adequately described in terms of the Eqs. (19), (20) and (21). However, it is possible to describe them in terms of the wave functions of the parent structure. For example, the skeletal molecular orbitals of a *nido* square pyramidal cluster can be expressed in terms of those of an octahedron. The correlation diagram for the skeletal molecular orbitals of an octahedron and a square pyramid are illustrated in Fig. 17. It can be seen that P_0^π, D_{2s}^π, $\bar{D}_{2s}^\pi$ and $\bar{P}_0^\pi$ are the same for both structures. Three *e* tangential molecular orbitals result from the mixings of $D_{\pm 1}^\pi$, $\bar{D}_{\pm 1}^\pi$, $P_{\pm 1}^\pi$ and $\bar{P}_{\pm 1}^\pi$ of the octahedron. The L^π and $\bar{L}^\pi$ molecular orbitals of an octahedron are illustrated in Fig. 18. The two π atomic orbitals at position 1 of an octahedron (π_1^θ, π_1^φ) can be expressed in terms of linear combinations of $D_{\pm 1}^\pi$, $\bar{D}_{\pm 1}^\pi$, $P_{\pm 1}^\pi$ and $\bar{P}_{\pm 1}^\pi$ using a localisation procedure developed by Mingos and Hawes [11]:

$$\pi_1^\theta = 1/2\,(P_{1c}^\pi + D_{1c}^\pi - \bar{P}_{1s}^\pi - \bar{D}_{1s}^\pi)$$

$$\pi_1^\varphi = 1/2\,(\bar{P}_{1c}^\pi + \bar{D}_{1c}^\pi + P_{1s}^\pi + D_{1s}^\pi)$$

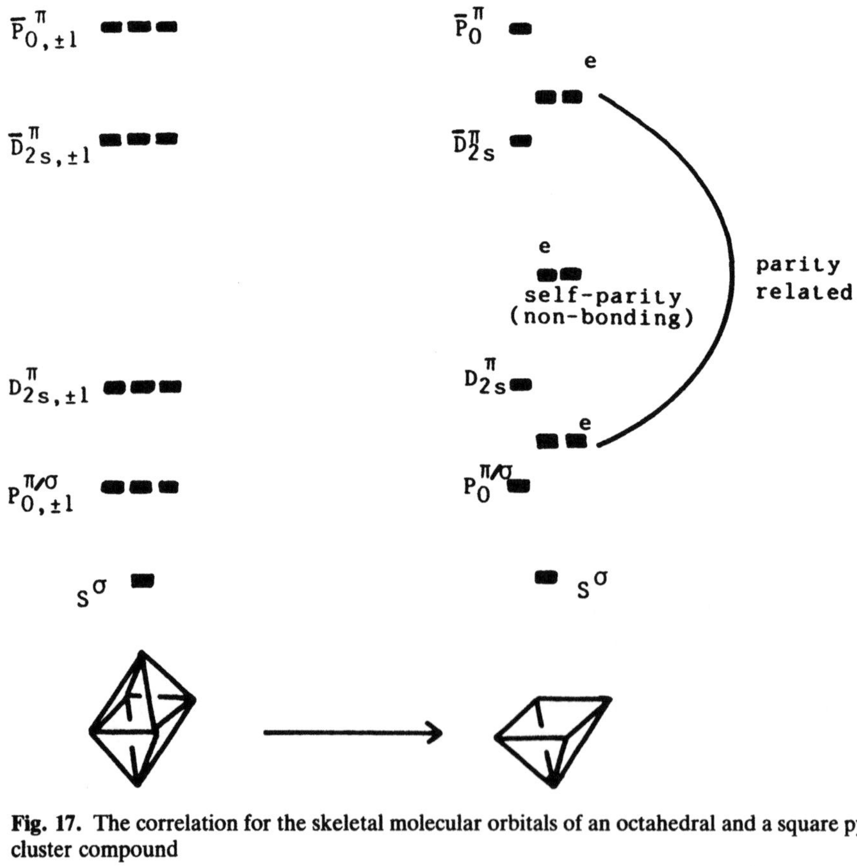

Fig. 17. The correlation for the skeletal molecular orbitals of an octahedral and a square pyramidal cluster compound

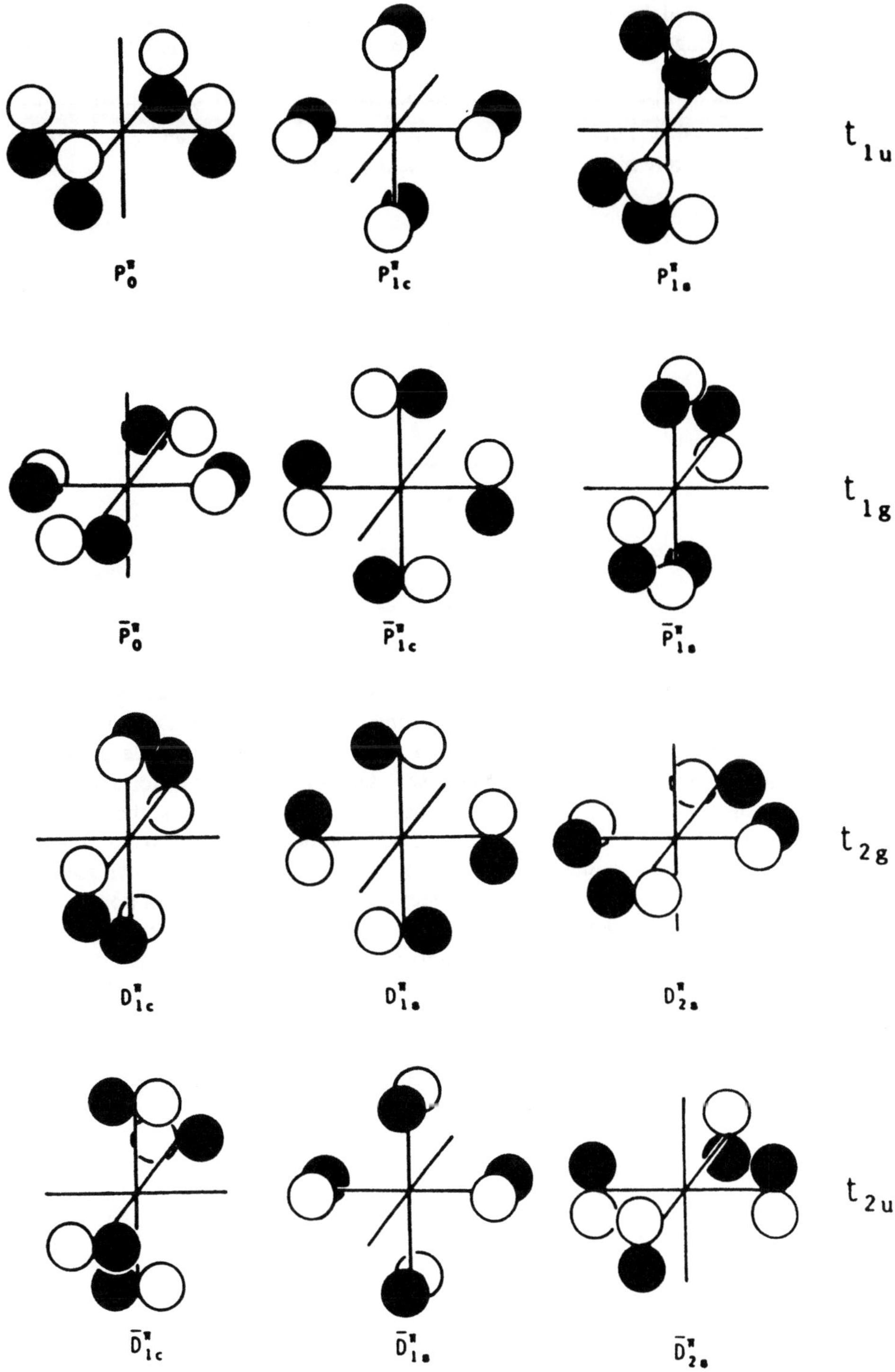

Fig. 18. The L^{π} and L^{π} molecular orbitals of an octahedral cluster

In the square pyramidal cluster, the interaction between the remaining three e sets gives rise to one *e* bonding, one *e* antibonding and one *e* non-bonding according to the **pairing principle.**

The expressions for π_1^θ and π_1^φ in terms of the octahedral $L^\pi/\bar{L}^\pi$ wave functions show that the π-tangential skeletal molecular orbitals of a *nido* structure can be described in terms of the TSH expressions for the parent *closo* structure.

(7)

In coordination compounds, ML_n, a non-bonding orbital that is a linear combination of atomic orbitals on the central atom was defined by using the quantum mechanical requirement that it has zero overlap with ligands. A similar method can be developed for the generation of the tangential non-bonding skeletal molecular orbitals for a *nido* cluster. In the pyramidal structure (**7**), one atom lies on the polar position and the remaining in the equatorial plane. Such a structure has an odd number of *e* irreducible representations and therefore a self-conjugate $L^\pi/\bar{L}^\pi$ non-bonding *e* set. Generally a *nido* structure is characterised by a total of n + 2 skeletal electron pairs, with the *e* set of non-bonding orbitals. Vector harmonics defining the *e* non-bonding orbitals in the pyramidal structure can be generated by using the following quantum mechanical requirements.

(1) This *e* set of non-bonding orbitals is a linear combination of $L^\pi_{1s,1c}$ and $\bar{L}^\pi_{1s,1c}$.

(2) Symmetry restrictions limit the mixing to occur only between L^π_{1s} and $\bar{L}^\pi_{1c}$, or L^π_{1c} and $\bar{L}^\pi_{1s}$.

(3) The mixing between L^π_{1s} and $\bar{L}^\pi_{1c}$ occurs in equal amount, i.e., the wave functions of the non-bonding orbitals have the form

$\sum_L c_L(L^\pi_{1c} + \bar{L}^\pi_{1s})$ because the energy of the non-bonding

orbital coincides with the mid-point of the energies of L^π and $\bar{L}^\pi$ in the Hückel approximation.

From Eqs. (20, 21) and the requirements above

$$\Psi^\pi_{\pm1}(\text{non-bonding}) = N \sum_i [V^\theta_{\pm1(\text{non})}(\theta_i, \varphi_i)\pi_i^\theta + V^\varphi_{\pm1(\text{non})}(\theta_i, \varphi_i)\pi_i^\varphi]$$

where

$$V^{\theta}_{\pm 1(\mathrm{non})} = c_1(p^{\theta}_{\pm 1} + \bar{p}^{\theta}_{\pm 1}) + c_2(d^{\theta}_{\pm 1} + \bar{d}^{\theta}_{\pm 1})$$

$$V^{\varphi}_{\pm 1(\mathrm{non})} = c_1(p^{\varphi}_{\pm 1} + \bar{p}^{\varphi}_{\pm 1}) + c_2(d^{\varphi}_{\pm 1} + \bar{d}^{\varphi}_{\pm 1})$$

In a real form,

$$\Psi^{\pi}_{1c,\,1s}(\text{non-bonding}) = N \sum_{i} [V^{\theta}_{1c,\,1s(\mathrm{non})}(\theta_i, \varphi_i)\pi^{\theta}_i + V^{\varphi}_{1c,\,1s(\mathrm{non})}(\theta_i, \varphi_i)\pi^{\varphi}_i] \quad (22)$$

and

$$V^{\theta}_{1s(\mathrm{non})} = c_1(p^{\theta}_{1s} + \bar{p}^{\theta}_{1c}) + c_2(d^{\theta}_{1s} + \bar{d}^{\theta}_{1c}) \quad (23)$$

$$V^{\varphi}_{1s(\mathrm{non})} = c_1(p^{\varphi}_{1s} + \bar{p}^{\varphi}_{1c}) + c_2(d^{\varphi}_{1s} + \bar{d}^{\varphi}_{1c})$$

The 1 c components can be obtained in a similar way. Inserting the wave functions of Table 6 into Eq. (23), we have:

$$V^{\theta}_{1s(\mathrm{non})} = [c_1(\cos\theta + 1) + c_2(3)^{1/2}(\cos 2\theta + \cos\theta)]\sin\varphi$$

$$V^{\varphi}_{1s(\mathrm{non})} = [c_1(\cos\theta + 1) + c_2(3)^{1/2}(\cos 2\theta + \cos\theta)]\cos\varphi$$

The wave function generated from the above vector harmonics (V^{θ}_{1s}, V^{φ}_{1s}) is the 1 s component of the e non-bonding orbital which has an orbital energy of α in a *nido* structure such as that shown in (7). This is true for all c_1/c_2 ratios because the L^{π} and $\bar{L}^{\pi}$ coefficients are equal. The vector can be expressed in the following form:

$[f(\theta)\sin\varphi, f(\theta)\cos\varphi]$

where $f(\theta) = c_1(\cos\theta + 1) + c_2(\cos 2\theta + \cos\theta)$. When θ is defined, e.g., for the equatorial ring, the vector is $c(\sin\varphi, \cos\varphi)$. The relevant vector diagram is illustrated in Fig. 19. Similarly the 1 c component is $[-f(\theta)\cos\varphi, f(\theta)\sin\varphi]$ which relates to the 1 s component by the parity relationship. Therefore, the 1 s and 1 c components are a pair of self-conjugate vector harmonics.

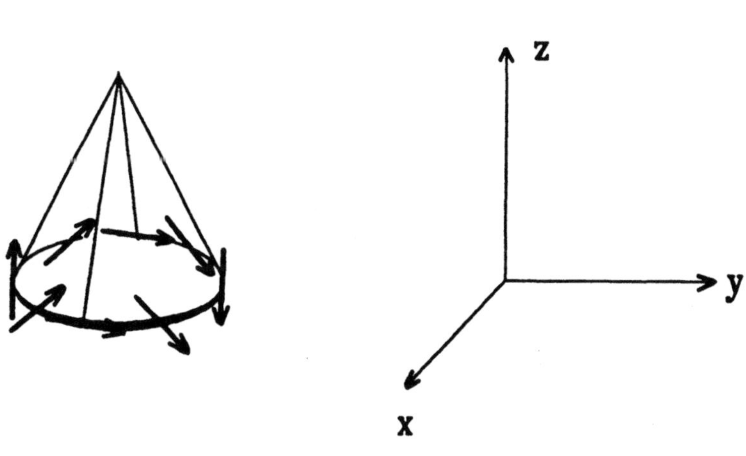

Fig. 19. The vector for describing the non-bonding orbital in a *nido* structure as shown in (7)

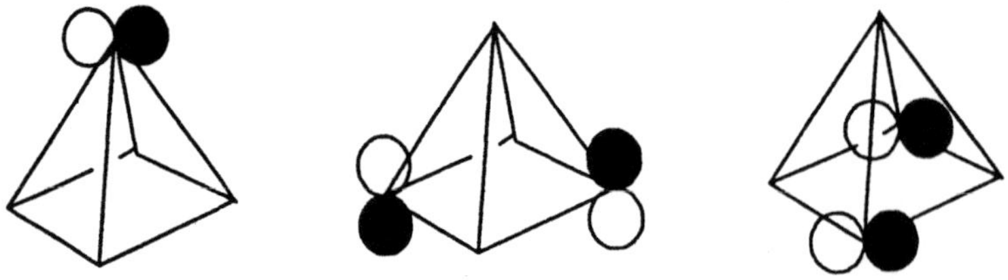

Fig. 20. The three e_{1s} components for a square pyramidal cluster

As the c_1/c_2 ratio is defined, the relative coefficients of the polar atom and the equatorial atoms can be determined. This can be achieved by defining some boundary conditions. The non-bonding orbital can be defined by considering the relevant secular determinant for the e representations. The three relevant e_{1s} components for a square pyramid are shown in Fig. 20. The secular determinant for the three e_{1s} components is:

$$\begin{array}{c} \\ e_{1s}^{(1)} \\ e_{1s}^{(2)} \\ e_{1s}^{(3)} \end{array} \begin{array}{c} \begin{array}{ccc} e_{1s}^{(1)} & e_{1s}^{(2)} & e_{1s}^{(3)} \end{array} \\ \left| \begin{array}{ccc} \alpha - E & \beta' & \beta' \\ \beta' & \alpha - E & 0 \\ \beta' & 0 & \alpha - E \end{array} \right| \end{array} = 0$$

where β' indicates the resonance integral between $e_{1s}^{(1)}$ and $e_{1s}^{(2)}$ (or $e_{1s}^{(1)}$ and $e_{1s}^{(3)}$, the assumption of $\beta^\sigma = \beta^\pi$ has been made). Comparing the above determinant with Eq. (1), it can be concluded that the non-bonding orbital with an orbital energy of α is localised on the equatorial atoms. Therefore, the assumption that $V_{1s(\text{non})}^{\theta} = V_{1s(\text{non})}^{\varphi} = 0$ when $\theta = 0$ can be made. However, when the number of atoms in the equatorial ring is increased the above assumption is invalid since the energies of the corresponding $e_{1s}^{(2)}$ and $e_{1s}^{(3)}$ components are not equal to α. The determinant now is:

$$\begin{array}{c} \\ e_{1s}^{(1)} \\ e_{1s}^{(2)} \\ e_{1s}^{(3)} \end{array} \begin{array}{c} \begin{array}{ccc} e_{1s}^{(1)} & e_{1s}^{(2)} & e_{1s}^{(3)} \end{array} \\ \left| \begin{array}{ccc} \alpha - E & \beta' & \beta' \\ \beta' & \alpha + \Delta - E & 0 \\ \beta' & 0 & \alpha - \Delta - E \end{array} \right| \end{array} = 0$$

For example, $\Delta = \beta$ for a hexagonal planar structure. The corresponding non-bonding orbitals are no longer localised on the equatorial atoms. Fortunately, the open face in most *nido* clusters contains 4 or 5 atoms. Therefore, the assumption can be taken as an approximate boundary condition. Under this assumption, $c_1/c_2 = -(3)^{1/2}$ and the following results are obtained:

$$V_{1s(\text{non})}^{\theta} = (1 - \cos 2\theta)\sin\varphi \qquad (24)$$

$$V_{1s(\text{non})}^{\varphi} = (1 - \cos 2\theta)\cos\varphi$$

In a similar way, the 1 c components are:

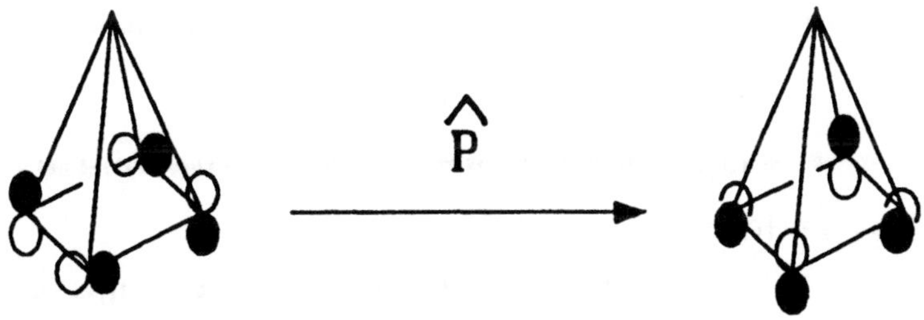

Fig. 21. The parity related *e* non-bonding orbitals in a square pyramidal cluster

$$V^{\theta}_{1c(non)} = -(1 - \cos 2\theta)\cos\varphi \tag{25}$$

$$V^{\varphi}_{1c(non)} = (1 - \cos 2\theta)\sin\varphi$$

The Eqs. (24) and (25) are a pair of self-conjugate vector harmonics. The vector harmonics ($V^{\theta}_{1s(non)}$, $V^{\varphi}_{1s(non)}$) in Eq. (24) are illustrated in Fig. 19. The 1 *c* component can be obtained by a rotation of the p^{π} orbitals in Fig. 19 through 90° about the + z axis. A specific example illustrating how the vector harmonics are used to define the non-bonding e_{1s} component for a square pyramid based on the vector diagram is illustrated in Fig. 21. The important point to emphasise with this *e* set is that the π^{θ} and π^{φ} components contribute equally and the parity relationship interconverts the e components. This ensures the non-bonding character of this molecular orbital. The effect of the parity relationship is illustrated in Fig. 21.

When the *nido* structure has three layers of atoms (**8**), the $f^{\pi}_{\pm 1}$ functions are involved in the non-bonding orbitals since the $F^{\pi}_{\pm 1}$ skeletal molecular orbitals are generated in the

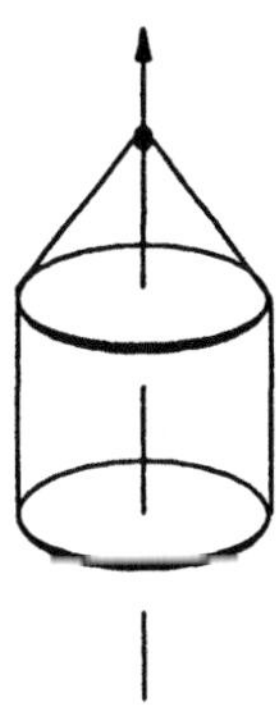

(**8**)

parent *closo* structure. In a similar way, the vector harmonics ($V^{\theta}_{1s(non)}$, $V^{\varphi}_{1s(non)}$) have the following linear combinations:

$$V^{\theta}_{1s(non)} = c_1(p^{\theta}_{1s} + \bar{p}^{\theta}_{1c}) + c_2(d^{\theta}_{1s} + \bar{d}^{\theta}_{1c}) + c_3(f^{\theta}_{1s} + \bar{f}^{\theta}_{1c})$$

$$V^{\varphi}_{1s(non)} = c_1(p^{\varphi}_{1s} + \bar{p}^{\varphi}_{1c}) + c_2(d^{\varphi}_{1s} + \bar{d}^{\varphi}_{1c}) + c_3(f^{\varphi}_{1s} + \bar{f}^{\varphi}_{1c})$$

Inserting the wave functions in Table 6 into above equations, we obtain:

$$V^{\theta}_{1s(non)} = \{c_1(\cos\theta + 1) + (3)^{1/2}c_2(\cos 2\,\theta + \cos\theta) + (3/8)^{1/2}c_3[\cos\theta(15\cos^2\theta - 11) + (5\cos^2\theta - 1)]\}\sin\varphi$$

$$V^{\varphi}_{1s(non)} = \{c_1(\cos\theta + 1) + (3)^{1/2}c_2(\cos 2\,\theta + \cos\theta) + (3/8)^{1/2}c_3[\cos\theta(15\cos^2\theta - 11) + (5\cos^2\theta - 1)]\}\cos\varphi$$

Again, the above vector can be expressed as $[f(\theta)\sin\varphi, f(\theta)\cos\varphi]$. Therefore, the vector in each ring has the same definition as that in Fig. 19. Consequently, the non-bonding character within each ring is retained. As above for a pyramidal structure, the assumption that $V^{\theta}_{1s(non)} = V^{\varphi}_{1s(non)} = 0$ when $\theta = 0$ is made again. Another condition is that when $\theta = 90°$ i.e., the equatorial plane between the two rings in (**8**), $V^{\theta}_{1s(non)} = V^{\varphi}_{1s(non)} = 0$. This implies that the vectors in the hemispheres above and below the equatorial plane change their signs. The sign changes ensure a zero overlap of the tangential orbitals between the two rings and retain their non-bonding character. The π^{θ} components have positive overlaps, whereas the π^{θ} components have negative overlaps. The total overlap is zero under the assumption of $\beta(\sigma) = \beta(\pi)$. This cancellation is illustrated schematically in Fig. 22. Therefore, we have the following equations:

$$c_1 + (3)^{1/2}c_2 + (6)^{1/2}c_3 = 0$$

$$c_1 - (3)^{1/2}c_2 - (6)^{1/2}c_3 = 0$$

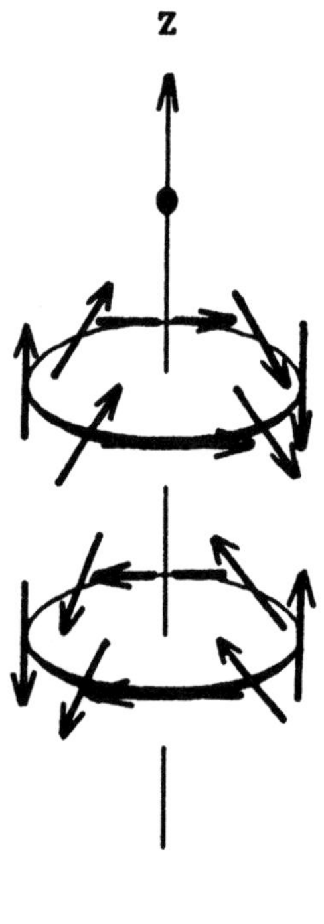

Fig. 22. The vector for describing the non-bonding orbital in a *nido* structure as shown in (**8**)

Solving the above equations, we obtain:

$$c_1 : c_2 : c_3 = (27/32)^{1/2} : (25/32)^{1/2} : (-1)$$

The final expressions for $V^{\theta}_{1s(non)}$ and $V^{\varphi}_{1s(non)}$ are:

$$V^{\theta}_{1s(non)} = \cos\theta \sin^2\theta \sin\varphi \qquad (26)$$

$$V^{\varphi}_{1s(non)} = \cos\theta \sin^2\theta \cos\varphi$$

The constant in front of the $\cos\theta$ term in the equation above has been ignored since it depends on the normalisation condition. The relevant vector diagram is illustrated in Fig. 22. The 1 c components have the following form:

$$V^{\theta}_{1c(non)} = -\cos\theta \sin^2\theta \cos\varphi \qquad (27)$$

$$V^{\varphi}_{1c(non)} = \cos\theta \sin^2\theta \sin\varphi$$

In summary, the non-bonding e orbitals for *nido* structures of the type illustrated in (**7**) or (**8**) can be obtained from Eqs. (22), (24) and (25) or Eqs. (22), (26) and (27).

An example illustrating the vector Eq. (26) to define the e_{1s} component of the e non-bonding orbitals for a capped-square antiprism is shown in Fig. 23. The non-bonding orbital has the following form according to Eq. (26):

$$\begin{aligned}\Psi_{(non)}(1\,s) = {} & 0.353(p_{\varphi 1} - p_{\varphi 3}) + 0.353(p_{\theta 2} - p_{\theta 4}) \\ & + 0.250[(-p_{\theta 5} - p_{\varphi 5}) + (-p_{\theta 6} + p_{\varphi 6}) + (p_{\theta 7} + p_{\varphi 7}) + (p_{\theta 8} - p_{\varphi 8})]\end{aligned}$$

From an extended Hückel molecular orbital calculation on $[B_9H_9]^{4-}$, this component is:

$$\begin{aligned}\Psi_{(non)}(1\,s) = {} & 0.120(p_{\varphi 1} - p_{\varphi 3}) + 0.230(p_{\theta 2} - p_{\theta 4}) \\ & + 0.275[(-p_{\theta 5} - p_{\varphi 5}) + (-p_{\theta 6} + p_{\varphi 6}) + (p_{\theta 7} + p_{\varphi 7}) + (p_{\theta 8} - p_{\varphi 8})] \\ & - 0.478\, p_{\varphi 9} \\ & + L^{\sigma} \text{ and H contributions}\end{aligned}$$

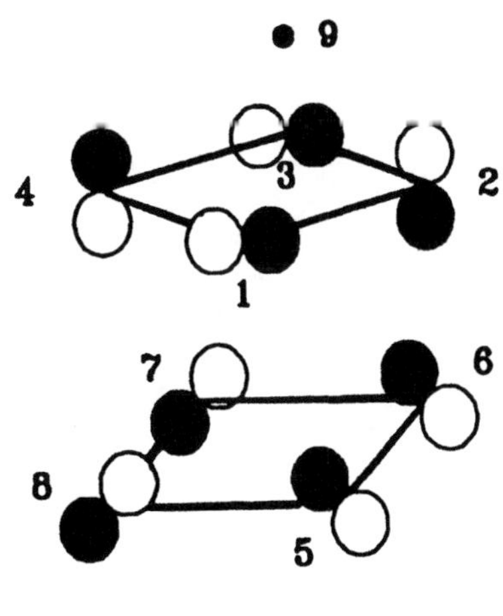

Fig. 23. The e_{1s} component of the e non-bonding orbitals in a capped square antiprismatic cluster

The above result indicates that there is a contribution from the polar atom in the non-bonding orbitals. This is due to a mixing between L^{σ} and L^{π}. It also means that the assumption of a zero vector in the polar position is not strictly valid. However, the sign of p_{θ} and p_{φ} is exactly matched between both results. Particular striking is the fact that the character contributed from those atoms in the open face is very close in both results.

The methodology developed above can also be used in those polar deltahedral clusters with $3p + 1$ vertices which have one *e* set of non-bonding orbitals. Consequently the vector diagram in Fig. 22 can be used to define the mixing between $L^{\pi}_{\pm 1}$ and $\bar{L}^{\pi}_{\pm 1}$ functions.

The mathematical aspects developed above can be summarised as follows. Whenever a *nido* polyhedron with C_{nv} ($n \geq 3$) symmetry is generated from a *closo* deltahedron it will have a non-bonding *e* set of molecular orbitals. This *e* set is localised predominately on the open face of the polyhedron and the π^{θ} and π^{φ} components conform to the vector diagrams shown in Figs. 19 and 22. If this *e* set is filled, then the *nido* polyhedron is characterised by n + 2 skeletal electron pairs.

The localisation of this *e* set on the open face of the polyhedron has two important consequences: firstly, they can be stabilised by overlap with the 1 *s* orbitals of protons leading to the *nido* boranes B_nH_{n+4}, and secondly they can be used as donor orbitals to transition metals to form *closo* metalloboranes.

If the *nido* polyhedron does not belong to a C_{nv} ($n \geq 3$) point group there will no longer be a non-bonding self-conjugate degenerate *e* set, but a pair of non-degenerate but parity related orbitals, which are approximately non-bonding. For example, for *nido* $[B_8H_8]^{4-}$, which is generated from a tricapped trigonal prism by the removal of a capping atom, there is a frontier orbital set of b_1 and b_2 symmetry which is illustrated in Fig. 24.

The relationship between *closo* $B_nH_n^{2-}$ which have spherical deltahedral structures and *nido* $B_nH_n^{4-}$ which have C_{nv} structures based on *closo* $B_{n+1}H_{n+1}^{2-}$ with a missing vertex is reminiscent of that noted previously for MX_n molecules with n electron pairs (spherical deltahedral) and those with n + 1 electron pairs which are based on MX_{n+1} polyhedra. In each case the more open C_{nv} structure results in the occupation of a non-bonding orbital in preference to an antibonding orbital in the parent spherical molecule. For MX_n, the choice of a C_{nv} structure is required by the necessity of creating a $s + ap_z + bd_{z^2}$ hybrid of a_1 symmetry which is noded in the M − X directions. For $B_nH_n^{4-}$ the C_{nv} geometry forces L^{π} and $\bar{L}^{\pi}$ components with *e* symmetry to mix to create an *e* pair of non-bonding tangential molecular orbitals. In the former case, there are an odd number of a_1 orbitals and, in the latter, an odd number of *e* representations in the $L^{\pi}/\bar{L}^{\pi}$ set.

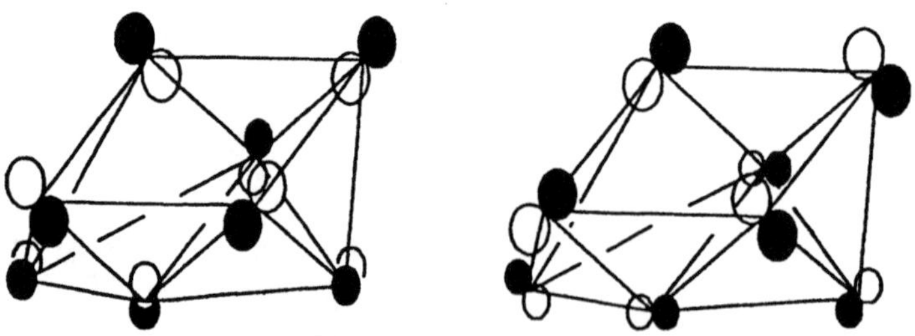

Fig. 24. The parity related b_1 and b_2 frontier orbitals for a bicapped trigonal prismatic cluster

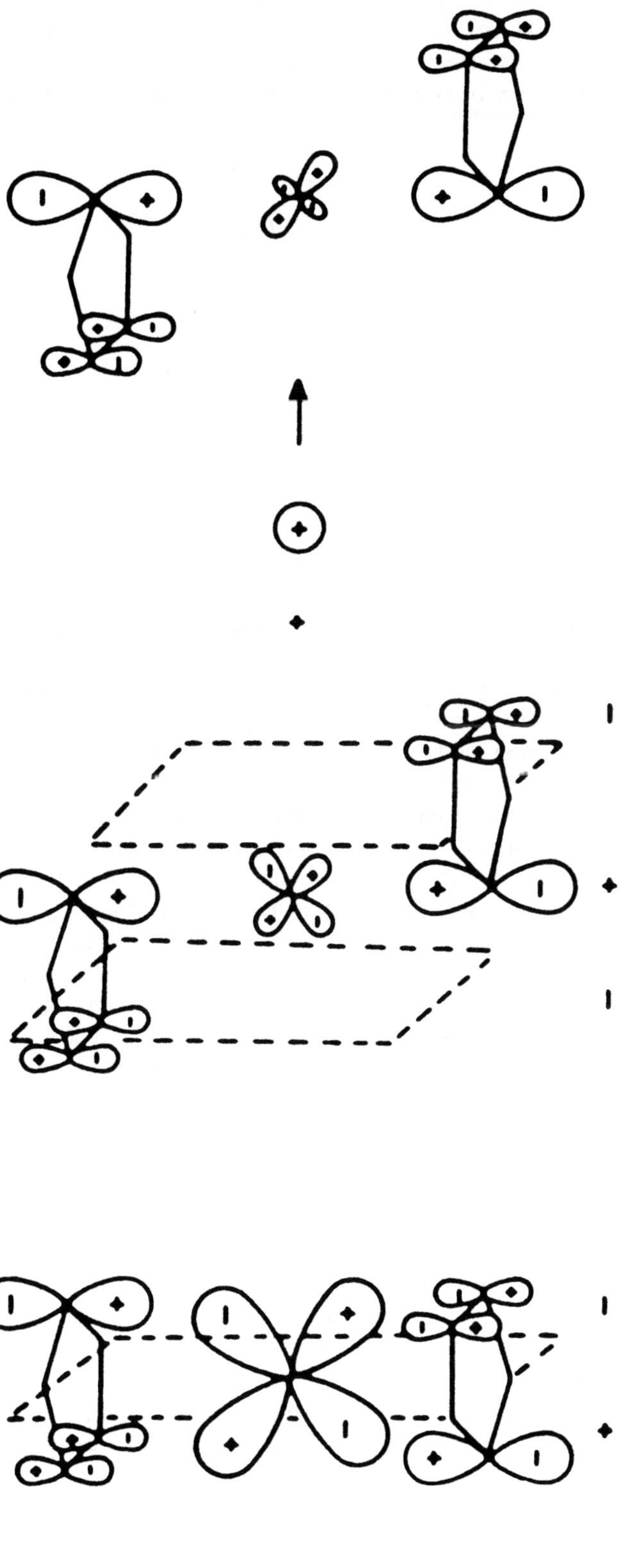

Fig. 25. The antibonding character of the $d_{xz} - e_1^{\pi}(x)$ molecular orbital in sandwich compounds and metallocarboranes

In sandwich compounds and metallocarboranes an alternative distortion mode is observed which converts an antibonding orbital into a non bonding orbital. This distortion has been described as a "slip" distortion [42–43] and involves the mutual displacement of the rings or *nido* polyhedra shown in (**9**). The manner in which the antibonding

(**9**)

character of the $d_{xz} - e_1^\pi(x)$ molecular orbital is reduced is illustrated schematically in Fig. 25. Particularly noteworthy is the manner in which the nodal cones of the metal and ligand orbitals move out of coincidence and thereby reduce their overlap.

Once again it should be emphasised that these distortions will only be observed when the initial orbital in the symmetrical structure is strongly antibonding. For example, first row metallocenes e.g., $Co(\eta\text{-}C_5H_5)_2$ [44] and $Ni(\eta\text{-}C_5H_5)_2$ [45] do not show such distortions, but second- and third-row examples where the metal-ring overlaps are larger do e.g., $Ru(\eta^6\text{-}C_6Me_6)(\eta^4\text{-}C_6Me_6)$ [46].

In an *arachno* structure generated by the removal of *trans* vertices of *closo* deltahedron the highest point group symmetry that can be achieved is D_{nd}. On each open face a non-bonding *e* set is generated by a removal of a single vertex. Therefore, the removal of two *trans* vertices simultaneously generates e_1 and $e_3(D_{4d})$ and e_{1g} and $e_{1u}(D_{5d})$ parity related and approximately non-bonding molecular orbitals. If these orbitals are fully occupied the *arachno* polyhedron is associated with n + 3 skeletal electron pairs. When the two open faces in an *arachno* structure are adjacent, the symmetry of the cluster is usually C_{2v}. The two *e* sets in fact transform as $2\ b_1 + 2\ b_2$. Therefore, the L^π and $\bar{L}^\pi$ molecular orbitals mix extensively but remain approximately non-bonding.

5 Summary

The discussion presented above has demonstrated that non-bonding orbitals play an important role in determining the stereochemistries and closed shell requirements of both co-ordination and cluster compounds. In group theoretical terms, whenever there are an odd number of irreducible representations, a non-bonding orbital is generated which is either localised exclusively on the central atom, or the peripheral atoms and has nodal characteristics which lead to a zero or minimal overlap with the remaining atoms in the structure.

The structure adopted by a molecule will always try and maximise the number of electron pairs in bonding molecular orbitals. A structure that is able to accommodate electron pairs in non-bonding orbitals will always be preferred to a more symmetrical structure which results in the population of antibonding molecular orbitals. Where more than one structure is possible with the same number of non-bonding orbitals, the one that maximises the character of the most stable atomic orbital is preferred. For example, both planar and pyramidal NH_3 have a non-bonding orbital but the latter has a higher percentage s character and, therefore, is preferred.

Acknowledgements. The SERC and the Chinese Academy of Sciences are thanked for their financial support and Drs. T. Slee and D. J. Wales for helpful comments.

6 References

1 Lewis GN (1966) Valence and the structure of atoms and molecules, Chemical Catalogue Co., New York (Reprinted by Dover, New York (1966))
2 Pauling L (1960) Nature of the chemical bond, 3rd Ed., Cornell University Press, Ithaca, New York
3 Pauling L (1975) Proc. Nat. Acad. Sci. USA. **72:** 4200 (1976) **73:** 274
4 Bent HA (1961) Chem. Rev. **61:** 275
5 Sidgwick NV, Powell HE (1940) Proc. Roy. Soc. London **A176:** 153
6 Gillespie RJ, Nyholm RS (1957) Quart. Rev. **11:** 339
7 Gillespie RJ (1972) Molecular geometry, Van Nostrand-Reinhold, Princeton, New Jersey
8 Ballhausen CJ, Gray HB (1964) Molecular orbital theory, W. A. Benjamin, New York
9 Mulliken RS (1955) J. Chem. Phys. **23:** 1833
10 Mingos DMP, Hawes JC (1985) J. Chem. Soc., Chem. Commun. 991
11 Mingos DMP, Hawes JC (1985) Structure and Bonding **63:** 1
12 Mingos DMP (1987) Pure & Appl. Chem. **59:** 145
13 Johnston RL, Mingos DMP (1989) Theor. Chim. Acta **75:** 11
14 Horn RA, Johnson, CR (1985) Matrix Analysis, Cambridge University Press
15 Rouvray DH, Balaban AT (ed) (1976) Chemical application of graphy theory, Ch 7, Academic, London
16 Gutman I, Trinajstic N (1973) Topics in Current Chemistry **42:** 49
17 Longuet-Higgins HC (1953) J. Chem. Phys. **18:** 265
Dewar MJS (1952) J. Amer. Chem. Soc. **74:** 3341
18 Cvetkovic D, Gutman I, Trinajstic N (1972) Croat. Chim. Acta. **44:** 365
19 Coulson CA, Rushbrooke S (1940) Proc. Camb. Phil. Soc. **36:** 193
20 Mingos DMP, Lin Zhenyang (1988) J. New. Chem. **12:** 787
21 Abrahams SC, Ginsberg AP, Knox K (1964) Inorg. Chem. **3:** 558
22 Keppert DL (1982) Inorganic stereochemistry, Springer, Berlin Heidelberg New York
23 Mutterties EL, Shunn RA (1966) Quart. Rev., Chem. Soc. **20:** 245
Mutterties EL, Guggenberger LJ (1974) J. Amer. Chem. Soc. **96:** 1748
24 Burdett JK (1980) Molecular shapes, John Wiley, New York
Albright TA, Burdett JK, Wangbo MH (1985) Orbital interactions in chemistry, John Wiley, New York
25 Lowe JP (1978) Quantum chemistry, Academic, New York
26 Ballard RE (1978) Photoelectron spectroscopy and molecular orbital theory, Adam Hilger, Bristol
27 Aung S, Pitzer RM, Chan SI (1968) J. Chem. Phys. **49:** 2071
28 Kutzelnigg W (1984) Angew. Chem. Int. Ed. Engl. **23:** 272
29 Mingos DMP, Lin Zhenyang (1989) Structure and Bonding (in press)

30 Ng J-W, Zuckerman JJ (1985) Adv. Inorg. Chem. Radiochem. **29:** 297
31 Woodward RB, Hoffmann R (1970) The conservation of orbital symmetry, Academic, New York
32 Zimmerman HE (1971) Acc. Chem. Res. **4:** 272
33 Wade K (1976) Adv. Inorg. Chem. Radiochem. **18:** 1
34 Mingos DMP (1986) Chem. Soc. Rev. **15:** 31
Mingos DMP, Johnston RL (1987) Structure and Bonding **68:** 29
35 Stone AJ (1980) Mol. Phys. **41:** 1339 (1981) Inorg. Chem. **20:** 256 (1984) Polyhedron **3:** 1299
36 Fowler PW (1985) Polyhedron **4:** 2051
37 Redmond DB, Quinn CM, McKiernan JGR (1983) J. Chem. Soc., Faraday Trans. 2, **79:** 1971
38 Johnston RL, Mingos DMP (1986) Polyhedron **5:** 2059
39 Eady CR, Johnson BFG, Lewis J, Hitchcock PB, Thomas KM (1977) J. Chem. Soc., Chem. Commun. 385
40 Bould J, Greenwood NN, Kennedy JD, McDonald WS (1982) J. Chem. Soc., Chem. Commun. 465
41 Johnston RL, Mingos DMP (1985) J. Organomet. Chem. **280:** 407
42 Mingos DMP, Forsyth MI (1978) J. Organomet. Chem. **146:** C37
Calhorda MJ, Mingos DMP, Welch AJ (1982) J. Organomet. Chem. **228:** 309
43 Mingos DMP (1977) J. Chem. Soc., Dalton Trans. 602
Mingos DMP, Forsyth MI, Welch AJ (1978) J. Chem. Soc., Dalton Trans. 1363
44 Bunder W, Weiss E (1975) J. Organomet. Chem. **92:** 65
45 Seiler P, Duâitz JD (1980) Acta. Cryst. **B36:** 2255
46 Huttner G, Lange S (1972) Acta. Cryst. **B28:** 2049

Vibronic Interactions in the Stereochemistry of Metal Complexes

Roman Boča, Martin Breza and Peter Pelikán

Departments of Inorganic and Physical Chemistry, Slovak Technical University,
CS-812 37 Bratislava, Czechoslovakia

A survey of the theoretical foundations of vibronic interactions in molecular systems, with special attention to metal complexes, is given. A detailed analysis of the conditions leading to the adiabatic potential surface in various degrees of generality represents the central idea of the article. Use of the partitioning method enables consideration of the Jahn-Teller effect, the Renner-Teller effect and the pseudo Jahn-Teller effect on an equal footing. Analytic forms of the adiabatic potential surfaces are rederived for the most important cases by including the totally symmetric vibrational mode up to the second order of vibronic expansion. A complete third-order formula for the adiabatic potential surface of $E_g - (a_{1g} + e_g)$ vibronic coupling is presented. The consequences for structural features of molecular systems are discussed and exemplified by various metal complexes. Recent progress in molecular-orbital calculations of vibronic constants is reviewed.

This paper deals with pure (pseudo) Jahn-Teller effect only. On the other hand, some qualitatively new types of interaction may arise due to interaction of (pseudo) Jahn-Teller centers with the lattice. We do not consider the competitive spin-orbit-lattice coupling that may arise below a magnetic-ordering temperature at Jahn-Teller ions with a threefold orbital degeneracy. Inclusion of interactions like this is, however, outside the scope of the present work.

1 Introduction 58
2 Separation of Nuclear Coordinates 60
3 Adiabatic Potential 63
4 Jahn-Teller Theorem 65
5 Analytic Form of the APS 70
6 Types of Vibronic Coupling 71
7 Approximations to $E - (a_1 + e)$ Vibronic Coupling 75
8 Symmetry Descent Concept 79
9 Experimental Structure Data 83
10 Molecular Orbital Calculations of Jahn-Teller Coupling Constants 86
11 Conclusions 90
12 Appendix. Reduction of Matrix Elements 90
13 Symbols and Abbreviations 95
14 References 96

© Springer-Verlag Berlin Heidelberg 1989

1 Introduction

The theoretical explanation of the existence of molecular structure is based on the concept of the adiabatic potential surface (APS) – a total energy function of fixed nuclear coordinates for molecular systems. As the APS links together molecular structure, spectroscopy, chemical thermodynamics and kinetics on a common basis, the methods predicting the shape of the APS belong to the most productive ideas of chemical physics[1,2].

In the usual approach, the APS is considered in a parametric form in which some constants, specifying its shape, occur. They may be obtained from experimental data and/or by theoretical calculations[3]. It must be mentioned that the APS concept originates in the Born-Oppenheimer (or adiabatic) approximation, where the nuclear motion is "frozen" (the nuclear mass being set infinite). However, the existence of electron degeneracy in certain nuclear configurations makes the problem of finding the analytic form of the APS more complex. Owing to electron – nuclear (vibronic) interactions, new phenomena appear among which the Jahn-Teller effect plays a key role. This has specific consequences on stereochemistry of metal complexes, and it brings rationalization of the distortion isomerism[4] (including cation and anion distortion isomerism, respectively[5]), the plasticity effect[6], fluxionality[7], considerable equatorial – axial interactions[8,9] or "chameleon behaviour"[10] of central atoms like Cu(II).

There is abundant literature concerning the Jahn-Teller effect and related phenomena; various aspects of this topic have been summed up in excellent reviews[11–16]. Unfortunately, many confusions, misunderstandings and inadequate terminology also can be met in some publications. In trying to rederive the basic equations for the APS in the presence of (pseudo) degeneracy, both the variation and perturbation methods may be used. It must be noted that they produce different results if nondegenerate, pseudodegenerate or strictly degenerate states are considered. This fact is of great importance; it is demonstrated by Table 1 for an interaction of two energy levels (see also Fig. 1). Evidently, a smooth transition must exist between the cases of degenerate and pseudodegenerate states, so that the Jahn-Teller effect and the pseudo Jahn-Teller effect should be explained on a common basis. Nevertheless, the former has been formulated in a truncated basis set of degenerate electronic states using the 1st-order perturbation theory. The latter, on the other hand, used to be explained by the 2nd-order perturbation theory (the so called second-order effect) in a complete (infinite) basis set of electronic

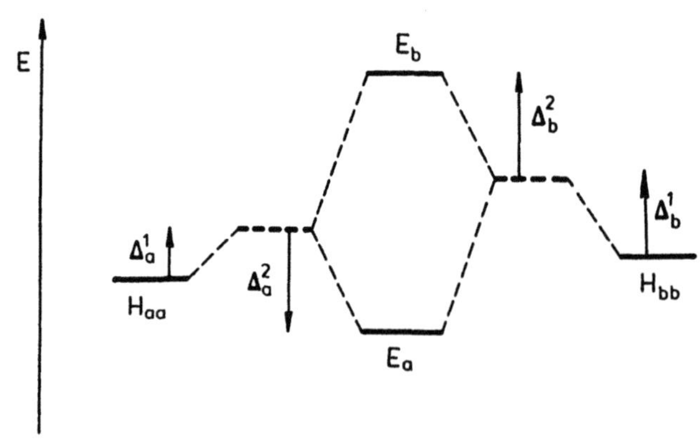

Fig. 1. Level shifting (Δ_a^1 and Δ_b^1) and level splitting (Δ_a^2 and Δ_b^2) by the interaction of two energy levels

Table 1. Results on two-level interaction by various methods[a]

	Variation method Nondegenerate	Degenerate	Perturbation method 1st-order Quasidegenerate	1st-order Degenerate	1st-order Degenerate	2nd-order Non-degenerate
Diagonal elements of the secular problem	H_{aa} H_{bb}	H_{aa} H_{aa}	$E_a^0 + H'_{aa}$ $E_b^0 + H'_{bb}$	$E^0 + H'_{aa}$ $E^0 + H'_{bb}$	$E^0 + H'_{aa}$ $E^0 + H'_{aa}$	$E_a^0 = H_{aa}^0$ $E_b^0 = H_{bb}^0$
Coefficients of quadratic secular equation	A: $1 - S^2$ B: $2H_{ab}S - (H_{aa} + H_{bb})$ C: $H_{aa}H_{bb} - H_{ab}^2$	$1 - S^2$ $2(H_{ab}S - H_{aa})$ $H_{aa}^2 - H_{ab}^2$	1 $-(E_a^0 + H'_{aa}) - (E_b^0 + H'_{bb})$ $(H'_{aa} + E_a^0)(H'_{bb} + E_b^0) - H'^2_{ab}$	1 $-(E^0 + H'_{aa}) - (E^0 + H'_{bb})$ $(H'_{aa} + E^0)(H'_{bb} + E^0) - H'^2_{ab}$	1 $-2(E^0 + H'_{aa})$ $(H'_{aa} + E^0)^2 - H'^2_{ab}$	
Discriminant $(D = B^2 - 4AC)$	D: $(H_{aa} - H_{bb})^2 + (H_{aa}S - H_{ab}) \times (H_{bb}S - H_{ab})$	$(H_{aa}S - H_{ab})^2$	$(E_a^0 + H'_{aa} - E_b^0 - H'_{bb})^2 + 4H'^2_{ab}$	$(H'_{aa} - H'_{bb})^2 + 4H'^2_{ab}$	$4H'^2_{ab}$	
Level shifting	Δ_a^1 $-B/2A$ Δ_b^1 $-B/2A$	$-B/2A$ $-B/2A$	$-B/2$ $-B/2$	$(H'_{aa} + H'_{bb})/2$ $(H'_{aa} + H'_{bb})/2$	H'_{aa} H'_{aa}	H'_{aa} H'_{bb}
Level splitting	Δ_a^2 $-\sqrt{D}/2A$ Δ_b^2 $+\sqrt{D}/2A$	$-\sqrt{D}/2A$ $+\sqrt{D}/2A$	$-\sqrt{D}/2$ $+\sqrt{D}/2$	$-\sqrt{D}/2$ $+\sqrt{D}/2$	$-\|H'_{ab}\|$ $+\|H'_{ab}\|$	$+H'^2_{ab}/\Delta_{ab}$ $-H'^2_{ab}/\Delta_{ab}$

[a] $S = \langle \phi_a \| \phi_b \rangle$, $H_{ab} = \langle \phi_a \| \hat{H} \| \phi_b \rangle$, $H_{ab}^0 = \langle \phi_a \| \hat{H}^0 \| \phi_b \rangle$, $H'_{ab} = \langle \phi_a \| \hat{H}' \| \phi_b \rangle$, $\hat{H} = \hat{H}^0 + \hat{H}'$, $\Delta_{ab} = E_a^0 - E_b^0$.

wave functions[17]. The variational formulation, restricting only to the first excited state of an appropriate symmetry, is also known[12]. Thus there is need for a more general formulation of vibronic coupling, and the approach outlined by Özkan and Goodman[18] seems to be an appropriate solution of the problem. The next section closely resembles their proposal based on the partitioning method.

Another problem originates in the fact that a multimode vibronic coupling should be considered in higher orders of the theory; at least the a_1 mode is to be included beginning with the second order. Therefore the analytic forms of the adiabatic potential surfaces need a rederivation. Finally, some structural transitions should be either included or deleted from the Jahn-Teller mechanism, and the symmetry descent concept is useful along these lines.

2 Separation of Nuclear Coordinates

Let us consider the Schrödinger equation for a molecule

$$H\psi = E\psi \,. \tag{1}$$

Its solution is of extreme difficulty; it consists of the following steps:

(i) construction of the molecular Hamiltonian;
(ii) expression of the molecular wave function in a suitable form;
(iii) application of the matrix partitioning method to reduce the dimensionality of the characteristic problem;
(iv) introduction of specific approximations that allow the solution of the problem.

The construction of the molecular Hamiltonian $\hat{H}$ is straightforward since it consists of the kinetic energy $\hat{T}$ of all particles (nuclei and electrons) and the potential energy $\hat{V}$ of all electrostatic interactions

$$\hat{H}(r, Q) = \hat{T} + \hat{V} = \hat{T}_N(Q) + \hat{T}_e(r) + \hat{V}_{NN}(Q) + \hat{V}_{ee}(r) + \hat{V}_{eN}(r, Q) \tag{2}$$

$\hat{T}_N$ being the nuclear kinetic energy, $\hat{T}_e$ – the kinetic energy of electrons, $\hat{V}_{NN}$ – the internuclear repulsion, $\hat{V}_{ee}$ – the interelectronic repulsion and $\hat{V}_{eN}$ – the electron-nuclear attraction.

The wave function ψ depends on all electron coordinates r_i and also on nuclear coordinates Q_j. Its exact expression in the Hilbert space is made possible by the expansion of

$$\psi(r, Q) = \sum_{k=0}^{\infty} f_k(Q)\, \phi_k(r; [Q^0]) \tag{3}$$

where the basis set is represented by electronic wave functions $\phi_k(r; [Q^0])$ for fixed nuclear coordinates Q_j^0, and the expansion coefficients $f_k(Q)$ are the nuclear (vibration) functions. The eletronic wave functions are orthonormal and they obey the electronic Schrödinger equation

$$\{\hat{T}_e + \hat{V} - E_k\}\phi_k(r; [Q]) = 0 \,. \tag{4}$$

However, this is not necessarily fulfilled since

$$H^{el}_{kl} = \int \phi^*_k(\hat{T}_e + \hat{V})\phi_l dr = \langle \phi_k|\hat{T}_e + \hat{V}|\phi_l\rangle_r = E^0_k\delta_{kl} + H'_{kl} \quad (5)$$

represents a matrix element of the electronic Hamiltonian $\hat{H}^{el} = \hat{T}_e + \hat{V}$ (it is non-zero, for example, for a couple of degenerate electronic states). Usually $[Q^0]$ is a configuration of nuclei in which $E_k[Q]$ displays a minimum value: $E_k[Q^0] = E^0_k$.

By substituting (3) into (1) we obtain

$$\sum_k \hat{H}|\phi_k\rangle f_k = E \sum_k |\phi_k\rangle f_k \quad (6)$$

which – after multiplication from the left side by $\langle\phi_l|$ and integration over electronic coordinates – yields the following set of differential equations

$$\sum_k \langle \phi_l|\hat{H}|\phi_k\rangle_r f_k = E f_l \quad \text{for } l = 1, 2, \ldots \quad (7)$$

where

$$\langle \phi_l|\hat{H}|\phi_k\rangle_r f_k = [\hat{T}_N(Q)\delta_{lk} + H^{el}_{lk} + \hat{G}_{lk}(Q)]f_k \quad (8)$$

The non-adiabatic coupling operator

$$\hat{G}_{lk}(Q) = \langle \phi_l|\hat{T}_N(Q)|\phi_k\rangle_r - \sum_A^N \frac{\hbar^2}{m_A} \langle \phi_l|\nabla_A|\phi_k\rangle_r \nabla_A \quad (9)$$

arises from the operations

$$\{\hat{T}_N(Q)\}(f\phi) = -\sum_A \frac{\hbar^2}{2m_A} \nabla^2_A(f\phi) = -\sum_A \frac{\hbar^2}{2m_A} (\phi\nabla^2_A f + 2\nabla_A\phi\nabla_A f + f\nabla^2_A\phi) . \quad (10)$$

Thus the exact Schrödinger equation for nuclear motion may be rewritten in the form

$$\{\hat{T}_N \mathit{1} + \hat{H} - E\mathit{1}\}f = 0 \quad (11)$$

with the interaction matrix elements

$$\hat{H}_{lk}(Q) = H^{el}_{lk} + \hat{G}_{lk}(Q) . \quad (12)$$

This system of equations is of infinite dimension; therefore it can be solved by the partitioning technique.

The complete space of solutions is split into two subspaces: the one S_a covers a limited number of solutions which we are interested in and the other S_b contains the excited states. Figure 2 shows three most important cases: non-degenerate, pseudodegenerate, and strictly degenerate states within the subspace S_a. With this partitioning Eq. (11) becomes

$$\begin{bmatrix} \hat{H}_{aa} + (\hat{T}_N - E)\mathit{1}_a ; & \hat{H}_{ab} \\ \hat{H}_{ba} & ; \hat{H}_{bb} + (\hat{T}_N - E)\mathit{1}_b \end{bmatrix} \begin{bmatrix} f_a \\ f_b \end{bmatrix} = 0 \quad (13)$$

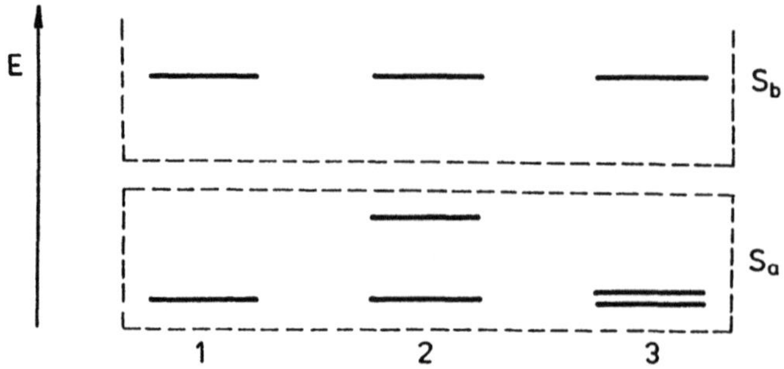

Fig. 2. Partitioning of states into strongly coupled subspace S_a and weakly coupled subspace of excited states S_b. *1* – nondegenerate state, *2* – pseudodegenerate states, *3* – strictly degenerate states

Here f_a and f_b are column vectors, $\mathit{1}_a$ and $\mathit{1}_b$ unit matrices and $\hat{H}_{aa}$, $\hat{H}_{ab}$, $\hat{H}_{ba}$ and $\hat{H}_{bb}$ are the submatrices of the given order. From the second equation for sub-matrices, one can express $f_b = \hat{R}_{bb}\hat{H}_{ba}f_a$ using the resolvent

$$\hat{R}_{bb} = (\mathrm{E}\mathit{1}_b - \hat{\mathrm{T}}_N\mathit{1}_b - \hat{H}_{bb})^{-1} . \tag{14}$$

Thus the solution of (13) may be found in the form of

$$\hat{H}^{\mathrm{eff}}_{aa}f_a = \mathrm{E}f_a \tag{15}$$

for the effective Hamiltonian in the subspace S_a of strongly coupled functions

$$\hat{H}^{\mathrm{eff}}_{aa} = \hat{\mathrm{T}}_N\mathit{1}_a + U_{aa} = \hat{\mathrm{T}}_N\mathit{1}_a + \hat{H}_{aa} + \hat{H}_{ab}\hat{R}_{bb}\hat{H}_{ba} . \tag{16}$$

Equation (15) is of principal significance since it is still the exact Schrödinger equation for nuclear motion in an applicable form.

In order to express the resolvent, the following formula should be applied to square matrices A and B

$$(A - B)^{-1} = A^{-1} + A^{-1}B(A - B)^{-1} = A^{-1} + A^{-1}BA^{-1} +$$

$$+ A^{-1}BA^{-1}B(A - B)^{-1} + \ldots = A^{-1}\sum_{p=0}^{\infty}(BA^{-1})^p . \tag{17}$$

Hence,

$$\hat{R}_{bb} = [(\mathrm{E}^0\mathit{1}_b - E^0_b) - (\hat{\mathrm{T}}_N\mathit{1}_b - \varepsilon\mathit{1}_b + H'_{bb} + \hat{G}_{bb})]^{-1} \tag{18}$$

may be written in various degrees of approximation, viz.

$$\hat{R}^{(0)}_{bb} = 0 \tag{19}$$

$$\hat{R}^{(1)}_{bb} = (\mathrm{E}^0\mathit{1}_b - E^0_b)^{-1} \tag{20}$$

$$\hat{R}^{(2)}_{bb} = \hat{R}^{(1)}_{bb} + (\mathrm{E}^0\mathit{1}_b - E^0_b)^{-1}(\hat{\mathrm{T}}_N\mathit{1}_b - \varepsilon\mathit{1}_b + H'_{bb} + \hat{G}_{bb})(\mathrm{E}^0\mathit{1}_b - E^0_b)^{-1} \tag{21}$$

where

$$\varepsilon = E - E^0 \tag{22}$$

is a pure vibration energy. The partitioning (18) guarantees that (20) is a good approximation to the resolvent since the second term in parentheses practically vanishes ($B \to 0$).

Also the electron interaction may be expressed in the form of a Taylor series

$$H^{el}_{kl} = E^0_k\delta_{kl} + H'_{kl} = E^0_k\delta_{kl} + H^{(1)}_{kl} + H^{(2)}_{kl} + \dots \tag{23}$$

where

$$H^{(1)}_{kl} = \sum_r \langle\phi^0_k| \left[\frac{\partial(\hat{T}_e + \hat{V})}{\partial Q_r}\right]_0 |\phi^0_l\rangle \Delta Q_r \tag{24}$$

$$H^{(2)}_{kl} = \frac{1}{2}\sum_{r,s} \langle\phi^0_k| \left[\frac{\partial^2(\hat{T}_e + \hat{V})}{\partial Q_r \partial Q_s}\right]_0 |\phi^0_l\rangle \Delta Q_r \Delta Q_s\ , \tag{25}$$

etc. The operator $\hat{T}_e$ may be omitted here since it is not a function of nuclear coordinates Q_r. For the same reason $\hat{V}$ becomes reduced to $\hat{V}_{NN} + \hat{V}_{eN}$ only. The summations run over all internal displacement coordinates ΔQ_r (usually symmetric coordinates measured relative to Q^0).

At this point, approximate expressions of (16) may be formulated. The usual treatment is to neglect the non-adiabatic coupling terms by setting $\hat{G}_{kl} \to 0$. Thus the interaction matrix elements $\hat{H}_{kl} \to H^{el}_{kl}$ represent the potential energy in the Schrödinger equation (11) for nuclear motion. In the limiting case of frozen nuclei ($\hat{T}_N \to 0$), the eigenvalues of H^{el}_{kl} become the total energy (because $\varepsilon \to 0$) and they represent the adiabatic potential surface.

3 Adiabatic Potential

Five important cases may be distinguished within the Born-Oppenheimer approximation (Table 2) in which the APS concept originates.

(i) The zeroth-order approximation to the resolvent (19) corresponds to the truncated basis set ($S_b \to 0$). If $\dim(S_a) = 1$, a non-degenerate state is under consideration. The adiabatic potential is simply $U_{kk} = E^0_k$, i.e. it is the solution of the electronic Schrödinger equation (4). In addition, if diagonal matrix elements $\hat{G}_{kk}(Q)$ are considered within the adiabatic approximation, the modified adiabatic potential $U_{kk} = E^0_k + \hat{G}_{kk}(Q)$ is obtained.

(ii) If $\dim(S_a) > 1$ within the truncated basis set and $E^0_k \approx E^0_l$, the case of pseudodegenerate states is considered. The adiabatic potential is obtained by solving the secular equation

Table 2. Origin of the adiabatic potential surface

Case	dim (S_a)	dim (S_b)	$R^{(n)}$	Interaction matrix elements	Note
1. Non-degenerate state	1	0	$R^{(0)}$	$U_{kk} = E_k^0$ $U_{kk} = E_k^0 + \hat{G}_{kk}(Q)$	Born-Oppenheimer approximation adiabatic approximation
2. Pseudodegenerate states within the truncated basis set	d	0	$R^{(0)}$	$U_{lk} = E_k^0\delta_{lk} + H_{lk}^{(1)}$	consistent with the variation method; pseudo Jahn-Teller effect
3. Degenerate states within the truncated basis set	d	0	$R^{(0)}$	$U_{lk} = E_0^0\delta_{lk} + H_{lk}^{(1)} + H_{lk}^{(2)}$	consistent with the first-order perturbation theory; Jahn-Teller or Renner-Teller effect
4. Ground state weakly coupled with excited states	1	∞	$R^{(1)}$	$U_{kk} = E_k^0 + \sum_{j \in S_b} H_{kj}^{(1)}H_{jk}^{(1)}/(E_k^0 - E_j^0)$	consistent with the second-order perturbation theory; pseudo Jahn-Teller effect
5. (Pseudo) degenerate states weakly coupled with excited states	d	∞	$R^{(1)}$	$U_{lk} = E_k^0\delta_{lk} + H_{lk}^{(1)} + H_{lk}^{(2)} + \sum_{j \in S_b} H_{lk}^{(1)}H_{jk}^{(1)}/(E_k^0 - E_j^0)$	general effect of vibronic coupling

$$\det\{U_{kl} - E\delta_{kl}\} = 0 \tag{26}$$

with

$$U_{kl} = E_k^0\delta_{kl} + H_{kl}^{(1)} + H_{kl}^{(2)} + \dots \tag{27}$$

This solution follows from application of the variational method.

(iii) If $E_k^0 = E_l^0$ holds within the truncated basis set, the strict degeneracy of electronic states occurs; usually twofold to fivefold degeneracy comes into consideration. The APS may again be obtained from Eq. (26).
This formulation is consistent with the result of the 1st-order perturbation theory for degenerate states.

(iv) If $\dim(S_a) = 1$ and the complete basis set is considered, i.e. $\dim(S_b) \rightarrow \infty$, weak coupling of the ground state with the excited states is taken into account. The resolvent in the form of (20) yields

$$U_{kk} = E_k^0 + E_{kk}^{(2)} = E_k^0 + \sum_{j \in S_b} H_{kj}^{(1)}H_{jk}^{(1)}/(E_k^0 - E_j^0) \, . \tag{28}$$

This formulation is consistent with the result of the 2nd-order perturbation theory for the energy (the 1st-order for the wave function).

(v) A more general case appears if $\dim(S_a) > 1$ and weak coupling with excited states is considered through $\hat{R}_{bb}^{(1)}$. Thus

$$U_{kl} = E_k^0\delta_{kl} + H_{kl}^{(1)} + H_{kl}^{(2)} + \dots + \sum_{j \in S_b} H_{kj}^{(1)}H_{jl}^{(1)}/(E_k^0 - E_j^0) \, . \tag{29}$$

This formula is applicable for both degenerate and pseudodegenerate states. The formulation of (29) allows consideration of the Jahn-Teller effect, the pseudo-Jahn-Teller effect and the Renner-Teller effect on a common basis in an arbitrary degree of approximation and the basis-set dimension.

4 Jahn-Teller Theorem

Let us consider the secular Eq. (26) in which the concept of the adiabatic potential originates. If the diagonal matrix elements are arranged in ascending order, $U_{11} < U_{22} \dots$, the non-zero off-diagonal elements cause at least one eigenvalue to fulfil the inequality $E_1 < U_{11}$. This result follows from the well known separation theorem[19]. A proof for the 2 × 2 problem is trivial since

$$2\,E_{1,2} = U_{11} + U_{22} \pm [(U_{11} - U_{22})^2 + 4\,U_{12}^2]^{1/2} \, . \tag{30}$$

Jahn and Teller have found[20] that it is valid for any non-linear molecule

$$H_{kl}^{(1)} = \sum_r \langle \phi_k^0 | \left[\frac{\partial(\hat{V}_{NN} + \hat{V}_{eN})}{\partial Q_r} \right]_0 | \phi_l^0 \rangle \Delta Q_r \neq 0 \tag{31}$$

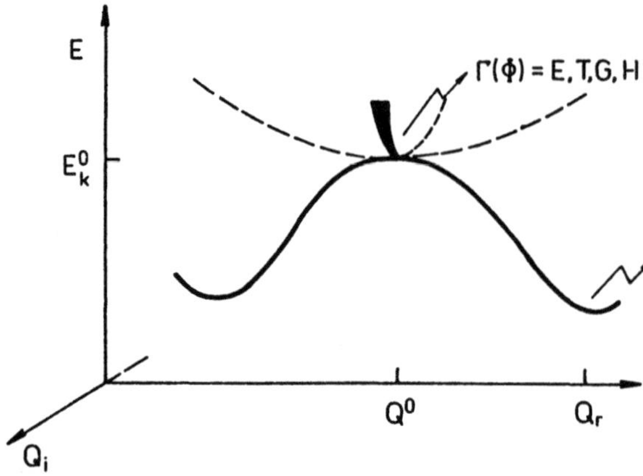

Fig. 3. Energy (symmetry) lowering along vibronic active coordinate Q_r

so that at least one vibrational mode Q_r exists for which the irreducible representation $\Gamma(Q_r)$ is contained in the symmetrized direct product of wave functions

$$[\Gamma_k \times \Gamma_l] = \Gamma_{red} = \Gamma(Q_r) + \ldots \tag{32}$$

(condition for the non-zero $\langle \Gamma_k | \Gamma(Q_r) | \Gamma_l \rangle$ type matrix element). This finding excludes the existence of an extremum at the point Q^0 of electron degeneracy (Fig. 3). Accordingly, the Jahn-Teller theorem may be formulated as:

(i) The electronic wave function corresponding to the minimum of the adiabatic potential cannot be orbitally degenerate (except in linear geometry); or
(ii) A non-linear configuration with a degenerate electronic state has a non-zero energy gradient and thus no energy minimum (it is unstable).

The theorem can be proved by looking at all point groups and their irreducible representations, taking into account the rules for the direct product of multidimensional irreducible representations[21]. The relevant rules are summarized in Table 3 together with the list of irreducible representations of the most important groups. A more general proof of the Jahn-Teller theorem can be found in Refs. 11, 22, 23.

Three effects based on the Jahn-Teller theorem can be distinguished (Fig. 4).

(i) The Jahn-Teller effect arises from the strict degeneracy at the reference geometry of a non-linear molecule as a consequence of $H_{kl}^{(1)} \neq 0$. Consequently, a symmetry descent with respect to Q^0 proceeds, and the electron degeneracy is removed. It has been termed the first-order effect although some second-order terms $H_{kl}^{(2)}$ adopt non-zero values; they are responsible for the stabilization of the respective geometry.
(ii) The Renner-Teller effect takes place for linear molecules where $H_{kl}^{(1)} = 0$. (It is a consequence of the fact that the distortion modes Σ^- and Π are absent within $[\Pi \times \Pi] = \Sigma^+ + \Delta$, etc.). As the $H_{kl}^{(2)}$ adopt a leading role, it has been termed the second-order effect. Depending on the value of $H_{kl}^{(2)}$, a bending of the linear molecule may occur.

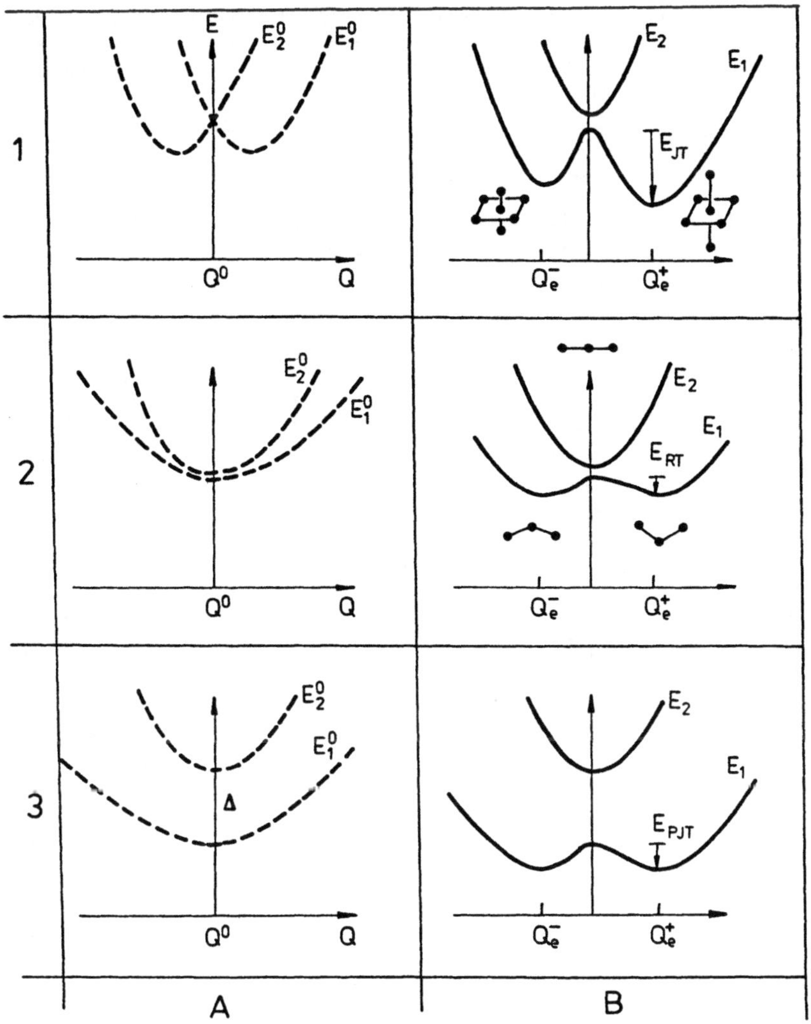

Fig. 4. Three effects of vibronic interactions. *1* – Jahn-Teller effect, *2* – Renner-Teller effect, *3* – pseudo-Jahn-Teller effect; **A** – before interaction, **B** – after interaction

(iii) The pseudo-Jahn-Teller effect occurs for pseudodegenerate states. Either

$$E_{kk'}^{(2)} = \sum_{j \in S_b} H_{kj}^{(1)} H_{jk'}^{(1)} / (E_k^0 - E_j^0) \neq 0 \tag{33}$$

or

$$H_{kl}^{(1)} \neq 0, E_k^0 < E_l^0 \tag{34}$$

are responsible for its existence. It has been termed the second-order effect, according to the use of the 2nd-order perturbation theory.

Table 3. Symmetrized direct products of irreducible representations and vibronic activity

Group	Degenerate electronic state	Vibronic active models[a, b]	Remaining modes[b]
Case 1. $[T \otimes T] = A_1 + E + T_2$			
O_h	$T_{1g}, T_{1u}, T_{2g}, T_{2u}$	A_{1g}, E_g, T_{2g}	$A_{1u}, A_{2g}, A_{2u}, E_u, T_{1g}, T_{1u}, T_{2u}$
O, T_d	T_1, T_2	A_1, E, T_2	A_2, T_1
T_h	T_g, T_u	$A_g, E_g, \underline{T_g}$	$A_u, E_u, \underline{T_g}, T_u$
T	T	$A, E, \underline{T}$	$\underline{T}$
Case 2. $[E \otimes E] = A_1 + B_1 + B_2$			
D_{2nh} $(n > 1)$	$E_{(n/2)g}, E_{(n/2)u}$	A_{1g}, B_{1g}, B_{2g}	$A_{1u}, A_{2g}, A_{2u}, B_{1u}, B_{2u}, E_{mg}, E_{mu}$ $(m < n)$
D_{2nd} $(n > 1)$	E_n	A_1, B_1, B_2	A_2, E_m $(m < 2n)$
D_{2n}, C_{2nv} $(n > 1)$	$E_{(n/2)}$	A_1, B_1, B_2	A_2, E_m $(m < n)$
C_{2nh} $(n > 1)$	$E_{(n/2)g}, E_{(n/2)u}$	$\underline{A_g}, B_g$	$\underline{A_g}, A_u, B_u, E_{mg}, E_{mu}$ $(m < n)$
C_{2n} $(n > 1)$	$E_{(n/2)}$	$\underline{A}, B$	$\underline{A}, E_m$ $(m < n)$
S_{4n}	E_n	$\underline{A}, B$	$\underline{A}, E_m$ $(m < 2n)$
Case 3. $[E \otimes E] = A_1 + E$			
O_h	E_g, E_u	A_{1g}, E_g	$A_{1u}, A_{2g}, A_{2u}, E_u, T_{1g}, T_{1u}, T_{2g}, T_{2u}$
O, T_d	E	A_1, E	A_2, T_1, T_2
T_h	E_g, E_u	$\underline{A_g}, E_g$	$\underline{A_g}, A_u, E_u, T_g, T_u$
T	E	$\underline{A}, E$	$\underline{A}, T$
D_{2nh} $(n > 2)$	E_{kg}, E_{ku} $\left(k < \frac{n}{2}\right)$	A_{1g}, E_{2kg}	$A_{1u}, A_{2g}, A_{2u}, B_{1g}, B_{1u}, B_{2g}, B_{2u}, E_{2ku}, E_{mg}, E_{mu}$ $(m \neq 2k, m < n)$
	E_{kg}, E_{ku} $\left(\frac{n}{2} < k < n\right)$	$A_{1g}, E_{(2n-2k)g}$	$A_{1u}, A_{2g}, A_{2u}, B_{1g}, B_{1u}, B_{2g}, B_{2u}, E_{(2n-2k)u}, E_{mg}, E_{mu}$ $(m \neq 2n-2k, m < n)$
$D_{(2n+1)h}$	E'_k, E''_k $\left(k < \frac{n}{2}\right)$	A'_1, E'_{2k}	$A''_1, A'_2, A''_2, E''_{2k}, E'_m, E''_m$ $(m \neq 2k, m \leqslant n)$
	E'_k, E''_k $\left(\frac{n}{2} \leqslant k \leqslant n\right)$	$A'_1, E'_{2n-2k+1}$	$A''_1, A'_2, A''_2, E''_{2n-2k+1}, E'_m, E''_m$ $(m \neq 2n-2k+1, m \leqslant n)$
D_{2nd} $(n > 2)$	E_k $(k < n)$	A_1, E_{2k}	A_2, B_1, B_2, E_m $(m \neq 2k, m < 2n)$
	E_k $(n < k < 2n)$	A_1, E_{4n-2k}	A_2, B_1, B_2, E_m $(m \neq 4n-2k, m < 2n)$
$D_{(2n+1)d}$	E_{kg}, E_{ku} $\left(k < \frac{n}{2}\right)$	A_{1g}, A_{2kg}	$A_{1u}, A_{2g}, A_{2u}, E_{2ku}, E_{mg}, E_{mu}$ $(m \neq 2k, m \leqslant n)$
	E_{kg}, E_{ku} $\left(\frac{n}{2} \leqslant k \leqslant n\right)$	$A_{1g}, E_{(2n-2k+1)g}$	$A_{1u}, A_{2g}, A_{2u}, E_{(2n-2k+1)u}, E_{mg}, E_{mu}$ $(m \neq 2n-2k+1, m \leqslant n)$
D_{2n}, C_{2nv} $(n > 1)$	E_k $\left(k < \frac{n}{2}\right)$	A_1, E_{2k}	A_2, B_1, B_2, E_m $(m \neq 2k, m < n)$
	E_k $\left(\frac{n}{2} < k < n\right)$	A_1, E_{2n-2k}	A_2, B_1, B_2, E_m $(m \neq 2n-2k, m < n)$

Table 3 (continued)

D_{2n+1}, $C_{(2n+1)v}$	$E_k \left(k < \frac{n}{2}\right)$	A_1, E_{2k}	A_2, E_m $(m \neq 2k, m \leq n)$
	$E_k \left(\frac{n}{2} \leq k \leq n\right)$	A_1, $E_{2n-2k+1}$	A_2, E_m $(m \neq 2n - 2k + 1, m \leq n)$
C_{2nh} $(n > 1)$	E_{kg}, E_{ku} $\left(k < \frac{n}{2}\right)$	$\underline{A}_g$, E_{2kg}	$\underline{A}_g$, A_u, B_g, B_u, E_{2ku}, E_{mg}, E_{mu} $(m \neq 2k, m < n)$
	E_{kg}, E_{ku} $\left(\frac{n}{2} < k < n\right)$	$\underline{A}_g$, $E_{(2n-2k)g}$	$\underline{A}_g$, A_u, B_g, B_u, $E_{(2n-2k)g}$, E_{mg}, E_{mu} $(m \neq 2n - 2k, m < n)$
$C_{(2n+1)h}$	E'_k, E''_k $\left(k < \frac{n}{2}\right)$	$\underline{A}'$, E'_{2k}	$\underline{A}'$, A'', E''_{2k}, E'_m, E''_m $(m \neq 2k, m \leq n)$
	E'_k, E''_k $\left(\frac{n}{2} \leq k \leq n\right)$	$\underline{A}'$, $E'_{2n-2k+1}$	$\underline{A}'$, A'', $E''_{2n-2k+1}$, E'_m, E''_m $(m \neq 2n - 2k + 1, m \leq n)$
C_{2n} $(n > 1)$	$E_k \left(k < \frac{n}{2}\right)$	$\underline{A}$, E_{2k}	$\underline{A}$, B, E_m $(m \neq 2k, m < n)$
	$E_k \left(\frac{n}{2} < k < n\right)$	$\underline{A}$, E_{2n-2k}	$\underline{A}$, B, E_m $(m \neq 2n - 2k, m < n)$
C_{2n+1}	$E_k \left(k < \frac{n}{2}\right)$	$\underline{A}$, E_{2k}	$\underline{A}$, E_m $(m \neq 2k, m < n)$
	$E_k \left(\frac{n}{2} < k < n\right)$	$\underline{A}$, $E_{2n-2k+1}$	$\underline{A}$, B, E_m $(m \neq 2n - 2k + 1, m \leq n)$
S_{4n}	E_k $(k < n)$	$\underline{A}$, E_{2k}	$\underline{A}$, B, E_m $(m \neq 2k, m < 2n)$
	E_k $(n < k < 2n)$	$\underline{A}$, E_{4n-2k}	$\underline{A}$, B, E_m $(m \neq 4n - 2k, m < 2n)$
S_{4n+2}	E_{kg}, E_{ku} $\left(k < \frac{n}{2}\right)$	$\underline{A}_g$, E_{2kg}	$\underline{A}_g$, B_u, E_{2ku}, E_{mg}, E_{mu} $(m \neq 2k, m \leq n)$
	E_{kg}, E_{ku} $\left(\frac{n}{2} \leq k \leq n\right)$	$\underline{A}_g$, $E_{(4n-2k+1)g}$	$\underline{A}_g$, B_u, $E_{(4n-2k+1)u}$, E_{mg}, E_{mu} $(m \neq 4n - 2k + 1, m \leq n)$
Case 4. $[G \otimes G] = A + G + H$			
I_h	G_g, G_u	A_g, G_g, H_g	A_u, T_{1u}, T_{1g}, T_{2u}, T_{2g}, G_u, H_u
I	G	A, G, H	T_1, T_2
Case 5. $[H \otimes H] = A + G + 2H$			
I_h	H_g, H_u	A_g, $\underline{G}_g$, H_g	A_u, T_{1g}, T_{1u}, T_{2g}, T_{2u}, $\underline{G}_g$, G_u, H_u
I	H	A, $\underline{G}$, H	T_1, T_2, $\underline{G}$
Case 6. The linear groups[c]			
$D_{\infty h}$	ε_{kg}, ε_{ku} $(k = 1, 2 \ldots)$	Σ_g^+, ε_{2kg}	Σ_g^-, Σ_u^+, Σ_u^-, ε_{2ku}, ε_{mg}, ε_{mu} $(m \neq 2k)$
$C_{\infty v}$	ε_k $(k = 1, 2 \ldots)$	Σ^+, ε_{2k}	Σ^-, ε_m $(m \neq 2k)$

[a] The vibronic activity in linear coupling is meant.

[b] The *underlined* modes are contained in symmetrized as well as in antisymmetrized direct products of irreducible representations of electronic states. There must be at least two sets of coordinates belonging to these modes and at least one of them is vibronic active.

[c] We denote: $\varepsilon_1 = \Pi$, $\varepsilon_2 = \Delta$, $\varepsilon_3 = \Phi$, etc.

5 Analytic Form of the APS

The usual procedure is to apply a Taylor expansion for the adiabatic potential surface

$$E(\dots Q_i \dots) = E^0 + K_1Q_1 + K_2Q_2 + \tfrac{1}{2}K_{11}Q_1^2 + K_{12}Q_1Q_2 + \tfrac{1}{2}K_{22}Q_2^2 + \dots \tag{35}$$

where the gradient components

$$K_i = \left(\frac{\partial E}{\partial Q_i}\right)_0 = \langle\phi_0| \left[\frac{\partial(\hat{V}_{NN} + \hat{V}_{eN})}{\partial Q_i}\right]_0 |\phi_0\rangle = F_i + X_i \tag{36}$$

vanish at the extremum point Q^0, so that the pure nuclear term, F_i, and the electron-nuclear term, X_i, are cancelled. The second derivatives

$$K_{ij} = \left(\frac{\partial^2 E}{\partial Q_i \partial Q_j}\right)_0 = \langle\phi_0| \left[\frac{\partial^2(\hat{V}_{NN} + \hat{V}_{eN})}{\partial Q_i \partial Q_j}\right]_0 |\phi_0\rangle + \left(\frac{\partial^2 E_{00}^{(2)}}{\partial Q_i \partial Q_j}\right)_0 \tag{37}$$

represent the harmonic force constants. The last term originates in the weak coupling with excited states and it gives a negative contribution to K_{ij}[17]. Higher-order force constants like K_{ijk} and K_{ijkl} describe the anharmonicity effects. Equation (35) is only applicable for the non-degenerate electronic states near the reference configuration Q^0. (A better description is obtained in relative coordinates $\varrho_i = (Q_i - Q_i^0)/Q_i^0$ or $\tau_i = (Q_i - Q_i^0)/Q_i$.)

The occurrence of electron degeneracy brings two effects. First, the molecular Hamiltonian is invariant with respect to arbitrary symmetry operations, so that the adiabatic potential belongs to the totally symmetric representation (A-type or Σ-type). Therefore, only certain combinations of Q_iQ_j or $Q_iQ_jQ_k$ are allowed for symmetry coordinates and they must span the totally symmetric representation. Secondly, the interaction matrix U

$$U_{kl} = E_k^0\delta_{kl} + H_{kl}^{(1)} + H_{kl}^{(2)} + \dots = E_k'\delta_{kl} + V_{kl} \tag{38}$$

has a scalar part

$$E_k' = E_k^0 + \sum_r K_rQ_r + \frac{1}{2}\sum_{r,s} K_{rs}Q_rQ_s + \dots \tag{39}$$

and a non-scalar part

$$V_{kl} = V_{kl}^{(1)} + V_{kl}^{(2)} + \dots = \sum_r \langle\phi_k^0| \left[\frac{\partial \hat{V}_{eN}}{\partial Q_r}\right]_0 |\phi_l^0\rangle \Delta Q_r$$

$$+ \frac{1}{2}\sum_{r,s} \langle\phi_k^0| \left[\frac{\partial^2 \hat{V}_{eN}}{\partial Q_r \partial Q_s}\right]_0 |\phi_l^0\rangle \Delta Q_r \Delta Q_s + \dots \tag{40}$$

The scalar part yields the usual (harmonic) Taylor expansion of the adiabatic potential, where the force constants occur. The non-scalar part may be reduced according to the Wigner-Eckart theorem to

$$V_{kl}^{r} = \langle \phi_k^0 | \left[\frac{\partial \hat{V}_{eN}}{\partial Q_r} \right]_0 | \phi_l^0 \rangle = X(\Gamma_k \Gamma_l; \Gamma_r) \left\langle \begin{matrix} \Gamma_l \Gamma_r \\ \gamma_l \gamma_r \end{matrix} \middle| \begin{matrix} \gamma_k \\ \Gamma_k \end{matrix} \right\rangle N \tag{41}$$

where $X(\Gamma_k\Gamma_l; \Gamma_r) = X_{\Gamma_r}$ is the reduced matrix element (considered as a parameter or the vibronic constant) and the Clebsh-Gordan coefficients have been tabulated (N is a conventional normalization factor). The reduction of a quadratic term, V_{kl}^{rs}, needs a successive decomposition of the operator part, $(\partial^2 \hat{V}_{eN}/\partial Q_r \partial Q_s)_0$, according to the Wigner formula

$$\left| \begin{matrix} \Gamma_r \Gamma_s \\ \gamma_r \gamma_s \end{matrix} \right\rangle = \sum_{\Gamma, \gamma} \begin{pmatrix} \Gamma_r \Gamma_s \Gamma \\ \gamma_r \gamma_s \gamma \end{pmatrix} |\Gamma_r \Gamma_s \Gamma_\gamma\rangle \tag{42}$$

where the 3 j-symbols (or Wigner coefficients) in parentheses also have been tabulated. The summations run over all Γ and γ contained in the reducible representation of the direct product $\Gamma_r \otimes \Gamma_s$.

Thus $V^{(2)}$ is a scalar convolution of two second-rank tensors, their components forming the basis of the $\Gamma_r \otimes \Gamma_s$ direct product. A more detailed explanation may be found in Appendix 1.

In this way, the non-scalar part forms a symmetric matrix *V;* its elements are a linear function of vibronic constants X and nuclear displacements ($Sp(V) = 0$). Its eigenvalues ε'

$$\det\{V_{kl} - \varepsilon' \delta_{kl}\} = 0 \tag{43}$$

represent a vibronic correction to the harmonic form of the adiabatic potential

$$E_k = E_k' + \varepsilon_k' \ . \tag{44}$$

This correction is responsible for the warping of the adiabatic potential surface near the nuclear configuration with electron degeneracy. The analytic form of the adiabatic potential surface becomes a non-linear (not only polynomial) function of nuclear displacements in which potential constants of two kinds occur. The force constants K_i, K_{ii}, K_{iik}, etc. contain the pure nuclear term F_i, F_{ii}, F_{iik}, etc. as well as the electron-nuclear (vibronic) term X_i, X_{ii}, X_{iik}, etc. The vibronic constants X_i, X_{ij}, etc. enter into the vibronic correction term ε'. They determine a warping of the adiabatic potential surface and they can couple vibration modes of different symmetry, e.g. X_{ij} for $i \neq j$.

6 Types of Vibronic Coupling

With the exception of icosahedral groups, only three basic types of vibronic coupling are possible (Table 3). They are:

(i) the three-mode coupling T-$(a_1 + e + t_2)$ for cubic groups (T_d, T_h, T, O_h and O) with triply degenerate electron terms;

(ii) the three-mode coupling E-$(a_1 + b_1 + b_2)$ for groups D_{4h}, C_{4v}, C_{4h}, C_4, S_4, D_{2d} and so on, where $E \otimes E = A_1 + A_2 + B_1 + B_2$ holds;

(iii) the two-mode coupling E-(a_1 + e) for the remaining groups with the doubly degenerate electron terms.

Usually the a_1 mode (the totally symmetric coordinate Q_1) is deleted from consideration since it has no influence in the linear approximation of $V_{kl}^{(1)}$. In quadratic approximation, however, this mode couples with other vibronic active modes.

In order to illustrate the behaviour of E-(a_1 + e) coupling, let us consider the displacements $Q_1(a_{1g})$, $Q_2(e_g)$ and $Q_3(e_g)$. From Appendix 1, it follows that

$$V = V^{(1)} + V^{(2)} = \frac{1}{\sqrt{2}}\begin{pmatrix}0 & 1\\ 1 & 0\end{pmatrix}(X_eQ_2 + X_{ae}Q_1Q_2 + X_{ee}Q_2Q_3/\sqrt{2})$$

$$+ \frac{1}{\sqrt{2}}\begin{pmatrix}-1 & 0\\ 0 & 1\end{pmatrix}[X_eQ_3 + X_{ae}Q_1Q_3 + X_{ee}(Q_2^2 - Q_3^2)/2\sqrt{2}] \,. \qquad (45)$$

The quadratic part

$$V^{(2)} = V^{(2)}[a + e, a + e] + V^{(2)}[t, t] + V^{(2)}[a + e, t]$$

$$= \frac{1}{2}\sum_{r=1}^{3}\sum_{s=1}^{3} V^{rs}[a + e, a + e]\Delta Q_r\Delta Q_s$$

$$+ \frac{1}{2}\sum_{r=4}^{15}\sum_{s=4}^{15} V^{rs}[t, t]\Delta Q_r\Delta Q_s$$

$$+ \frac{1}{2}\sum_{r=1}^{3}\sum_{s=4}^{15} V^{rs}[a + e, t]\Delta Q_r\Delta Q_s \qquad (46)$$

includes the first term only. The second term is not considered within the E_g-(a_{1g} + e_g) coupling model. The last term vanishes exactly by the symmetry. In this way, the number of 2 × 15 × 15 × 2 = 900 quadratic integrals, V_{kl}^{rs}, becomes reduced to 16 non-zero elements.

Since the potential matrix U adopts the form of

$$U = E^0 1 + H^{(1)} + H^{(2)} = E' 1 + V^{(1)} + V^{(2)} = c_1C_1 + c_2C_2 + c_3C_3 \qquad (47)$$

its eigenvalues are

$$E = c_1 \pm [(c_2^2 + c_3^2)/2]^{1/2} \qquad (48)$$

The vibronic correction to the adiabatic potential yields

$$\varepsilon' = \pm [(X_e + X_{ae}Q_1)^2(Q_2^2 + Q_3^2)/2 + X_{ee}^2(Q_2^2 + Q_3^2)/16$$

$$+ X_{ee}(X_e + X_{ae}Q_1)Q_3(3\,Q_2^2 - Q_3^2)/2\sqrt{2}]^{1/2} \,. \qquad (49)$$

With the polar coordinates

$$\varrho = (Q_2^2 + Q_3^2)^{1/2},\ \varphi = \mathrm{arctg}(Q_2/Q_3) \qquad (50)$$

and the substitutions

$$A = X_e/\sqrt{2},\ B = X_{ee}/4,\ Z = X_{ae}/\sqrt{2} \tag{51}$$

for vibronic constants, the analytic form of the APS is

$$E_1(Q_1, \varrho, \varphi) = E^0 + \tfrac{1}{2} K_{aa}Q_1^2 + \tfrac{1}{2} K_{ee}\varrho^2 \pm \varrho[(A + ZQ_1)^2 + B^2\varrho^2 - 2(A + ZQ_1)B\varrho \cos(3\varphi)]^{1/2} . \tag{52}$$

In graphic representation, a "Mexican-hat" potential with warping is obtained (Fig. 5).

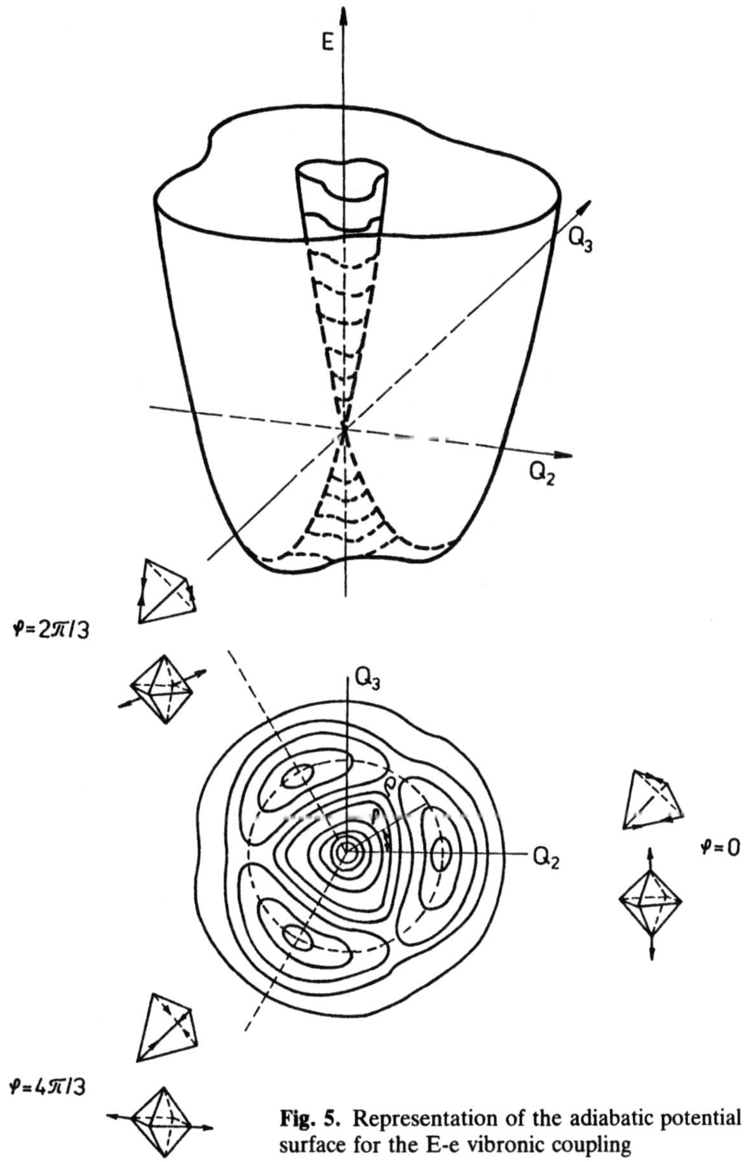

Fig. 5. Representation of the adiabatic potential surface for the E-e vibronic coupling

This result is valid for an arbitrary molecular system with E-(a_1 + e) coupling. Important examples are:

(i) the octahedron (O_h) with tetragonal distortions;
(ii) the tetrahedron (T_d);
(iii) the cube of O_h symmetry;
(iv) the triangle of D_{3h} symmetry;
(v) the antiprism of D_{3d} symmetry;
(vi) the linear system with $X_e = 0$ (Renner-Teller effect) where $\varepsilon' = \pm |X_{ee}|\varrho^2$.

Since the above potential (52) includes the vibrational part to the second order and the vibronic corrections to the second order as well, it is called the [2/2]-potential within the E-(a_1 + e) coupling.

The second type of vibronic coupling E-($a_1 + b_1 + b_2$) is characterized by $Q_1(a_{1g})$, $Q_2(b_1)$ and $Q_3(b_2)$ modes and it yields

$$V = \frac{1}{\sqrt{2}}\begin{pmatrix} 0 & 1 \\ 1 & 0 \end{pmatrix}[(X_2 + X_{12}Q_1)Q_2] + \frac{1}{\sqrt{2}}\begin{pmatrix} -1 & 0 \\ 0 & 1 \end{pmatrix}[(X_3 + X_{13}Q_1)Q_3] \tag{53}$$

so that the vibronic correction is[25)]

$$\varepsilon' = \pm [(X_2 + X_{12}Q_1)^2 Q_2^2/2 + (X_3 + X_{13}Q_1)^2 Q_3^2/2]^{1/2} \,. \tag{54}$$

Finally, the last type of vibronic coupling is T-(a_1 + e + t_2); it operates over six coordinates $Q_1(a_{1g})$, Q_2 and $Q_3(e_g)$, Q_4, Q_5 and $Q_6(t_{2g})$ for the octahedral reference configuration. The corresponding matrix elements V_{ij} are listed in Table 4 and the vibronic corrections to the APS ε' are obtained as eigenvalues of this 3 × 3 matrix.

Table 4 summarizes the expressions for the vibronic matrix elements V_{ij}, which all correspond to the [2/2]-type of the adiabatic potential.

Table 4. Types of vibronic coupling

Type	Active vibrations	$V_{kl} =$	V_{lk}
1. T-(a_1 + e + t_2)	$Q_1(a_{1g})$, $Q_2(e_g)$, $Q_3(e_g)$, $Q_4(t_{2g})$, $Q_5(t_{2g})$, $Q_6(t_{2g})$	$V_{11} =$	$X_e(Q_3 - \sqrt{3}Q_2)/2 + X_{ee}(Q_2^2 - Q_3^2 - 2\sqrt{3}Q_2Q_3)/2 + X_{ae}Q_1(Q_3 - \sqrt{3}3Q_2)/2 + X_{tt}(-2Q_4^2 + Q_5^2 + Q_6^2)/\sqrt{6}$
		$V_{12} =$	$X_tQ_6 - X_{et}Q_3Q_6 + X_{tt}Q_4Q_5 + X_{at}Q_1Q_6$
		$V_{13} =$	$X_tQ_5 + X_{et}Q_5(Q_3 + \sqrt{3}Q_2)/2 + X_{at}Q_1Q_5 + X_{tt}Q_4Q_6$
		$V_{22} =$	$X_e(Q_3 + \sqrt{3}Q_2) + X_{ee}(Q_2^2 - Q_3^2 + 2\sqrt{3}Q_2Q_3)/2 + X_{ae}Q_1(Q_3 + \sqrt{3}3Q_2)/2 + X_{tt}(Q_4^2 - 2Q_5^2 + Q_6^2)/\sqrt{6}$
		$V_{23} =$	$X_tQ_4 + X_{et}Q_4(Q_3 - \sqrt{3}Q_2)/2 + X_{at}Q_1Q_4 + X_{tt}Q_5Q_6$
		$V_{33} =$	$-X_eQ_3 - X_{ee}(Q_2^2 - Q_3^2) + X_{tt}(Q_4^2 + Q_5^2 - 2Q_6^2)/\sqrt{6} - X_{ae}Q_1Q_3$
2. E-($a_1 + b_1 + b_2$)	$Q_1(a_{1g})$, $Q_2(b_1)$, $Q_3(b_g)$	$-V_{11} =$	$V_{22} = (X_3 - X_{13}Q_1)Q_3/\sqrt{2}$
		$V_{12} =$	$(X_2 - X_{12}Q_1)Q_2/\sqrt{2}$
3. E-(a_1 + e)	$Q_1(a_{1g})$, $Q_2(e_g)$, $Q_3(e_g)$	$-V_{11} =$	$V_{22} = [X_{ae}Q_1Q_3 + X_{ee}(Q_2^2 - Q_3^2)/2\sqrt{2}]/\sqrt{2} + X_eQ_3/\sqrt{2}$
		$V_{12} =$	$(X_{ae}Q_1Q_2 + X_{ee}Q_2Q_3/\sqrt{2})/\sqrt{2} + X_eQ_2/\sqrt{2}$

7 Approximations to E-(a_1 + e) Vibronic Coupling

Works dealing with the static Jahn-Teller effect usually are based on the adiabatic potential in which only the single mode E-e coupling is considered. Within this more approximate approach, the coupling constant X_{ae} is omitted and the [2/2] potential is simplified to

$$E = E^0 + \tfrac{1}{2} K_{aa}Q_1^2 + \tfrac{1}{2} K_{ee}\varrho^2 \pm \varrho[A^2 + B^2\varrho^2 - 2\,AB\varrho\cos(3\,\varphi)]^{1/2}\,. \tag{55}$$

The ways of determining the potential parameters K_{aa}, K_{ee}, A and B, as well as the Jahn-Teller radius ϱ_0, are discussed elsewhere[3, 11, 12, 16]. The last formula differs from (52) only by the substitution of A + ZQ_1 for A. Notice that Z ~ X_{ae} should be considered as an observable that differs from B ~ X_{ee} (unlike the assumption of Coffman[27]).

Several authors[16, 28, 29] have tried to introduce a third-order term

$$K^{(3)} = K_{eee}(3\,Q_2^2Q_3 - Q_3^3) = -\,K_{eee}\varrho^3\cos(3\,\varphi) \tag{56}$$

where the constant K_{eee} has a dimension of energy per volume. This term is also responsible for warping the "Mexican hat"-type potential. The corresponding [3/1] potential

$$E = E^0 + \tfrac{1}{2} K_{ee}\varrho^2 \pm A\varrho - K_{eee}\varrho^3\cos(3\,\varphi) \tag{57}$$

has extreme points (three absolute minima and three saddle points) for angular coordinates $\varphi = n\pi/3$, which are analogous to Eq. (55).

A more complex [3/2] potential within the E-e coupling model was considered by Goodenough[30] in the form of

$$E = E^0 + \tfrac{1}{2} K_{ee}\varrho^2 - K_{eee}\varrho^3\cos(3\,\varphi) \pm \varrho[A^2 + B^2\varrho^2 - 2\,AB\varrho\cos(3\,\varphi)]^{1/2}\,. \tag{58}$$

Its counterpart within the more sophisticated E-(a_1 + e) coupling scheme is

$$\begin{aligned} E = {} & E^0 + \tfrac{1}{2} K_{aa}Q_1^2 + \tfrac{1}{2} K_{ee}\varrho^2 + K_{aaa}Q_1^3 + K_{aee}Q_1\varrho^2 - K_{eee}\varrho^3\cos(3\,\varphi) \\ & \pm \varrho[(A + ZQ_1)^2 + B^2\varrho^2 - 2\,(A + ZQ_1)B\varrho\cos(3\,\varphi)]^{1/2}\,. \end{aligned} \tag{59}$$

The necessity to include a cubic term $K^{(3)}$ was discussed elsewhere[31]. Only certain combinations of K_{ijk} are possible since the molecular Hamiltonian spans the totally symmetric representation (a_1 or σ^+) of the symmetry point group G_0. The method outlined by Goodenough[30] may be used in determining $K^{(3)}$ or higher-order terms.

Finally, Englman[11] presented the third-order vibronic correction in the form of

$$V^{(3)} = \frac{\sqrt{6}}{8}\,(Q_2^2 + Q_3^2)\begin{pmatrix} -Q_3 & Q_2 \\ Q_2 & Q_3 \end{pmatrix} X_{eee}\,. \tag{60}$$

The corresponding [3/3] potential within the E-e coupling scheme is

$$\begin{aligned} E = {} & E^0 + \tfrac{1}{2} K_{ee}\varrho^2 - K_{eee}\varrho^3\cos(3\,\varphi) \\ & \pm \varrho[(A + C\varrho^2)^2 + B^2\varrho^2 - 2\,(A + C\varrho^2)B\varrho\cos(3\,\varphi)]^{1/2} \end{aligned} \tag{61}$$

where $C = X_{eee}\sqrt{6}/8$. No additional warping is obtained by the third-order vibronic correction, and a formal analogy with (59) is evident.

With the results of Appendix 1, a complete third-order vibronic term can be obtained

$$\begin{aligned} V^{(3)} &= C_2\left\{\tfrac{1}{2} X_{aae}Q_1^2Q_2 + \frac{1}{\sqrt{2}} X_{aee}Q_1Q_2Q_3 + \frac{\sqrt{2}+1}{12} X_{eee}Q_2(Q_2^2 + Q_3^2)\right\} \\ &+ C_3\left\{\tfrac{1}{2} X_{aae}Q_1^2Q_3 + \frac{1}{2\sqrt{2}} X_{aee}Q_1(Q_2^2 - Q_3^2) + \frac{\sqrt{2}+1}{12} X_{eee}Q_3(Q_2^2 + Q_3^2)\right\} . \end{aligned} \tag{62}$$

Thus the vibronic matrix becomes

$$\begin{aligned} V &= V^{(1)} + V^{(2)} + V^{(3)} \\ &= C_2\sqrt{2}\,[A_1Q_2 + 2\,A_2Q_2Q_3] + C_3\sqrt{2}\,[A_1Q_3 + A_2(Q_2^2 - Q_3^2)] \\ &= A_1\begin{pmatrix} -Q_3 & Q_2 \\ Q_2 & Q_3 \end{pmatrix} + A_2\begin{pmatrix} -(Q_2^2 - Q_3^2)\;; & 2\,Q_2Q_3 \\ 2\,Q_2Q_3 & ;\;(Q_2^2 - Q_3^2) \end{pmatrix} \end{aligned} \tag{63}$$

with

$$A_1 = \frac{1}{\sqrt{2}}\left[X_e + X_{ae}Q_1 + \tfrac{1}{2} X_{aae}Q_1^2 + \frac{\sqrt{2}+1}{12} X_{eee}(Q_2^2 + Q_3^2)\right] \tag{64}$$

$$A_2 = \frac{1}{2\sqrt{2}}\left[\frac{1}{\sqrt{2}} X_{ee} + \frac{1}{\sqrt{2}} X_{aee}Q_1\right] . \tag{65}$$

The vibronic correction is

$$\begin{aligned} \varepsilon' &= \pm\{(A_1Q_2 + 2\,A_2Q_2Q_3)^2 + [A_1Q_3 + A_2(Q_2^2 - Q_3^2)]^2\}^{1/2} \\ &= \pm\,\varrho[A_1^2 + A_2^2\varrho^2 - 2\,A_1A_2\varrho\cos(3\,\varphi)]^{1/2} \end{aligned} \tag{66}$$

so that the adiabatic potential adopts the final form of

$$\begin{aligned} E &= E^0 + \tfrac{1}{2} K_{aa}Q_1^2 + \tfrac{1}{2} K_{ee}\varrho^2 + K_{aaa}Q_1^3 + K_{aee}Q_1\varrho^2 - K_{eee}\varrho^3\cos(3\,\varphi) \\ &\quad \pm\,\varrho[A_1^2 + A_2^2\varrho^2 - 2\,A_1A_2\varrho\cos(3\,\varphi)]^{1/2} . \end{aligned} \tag{67}$$

The last expression represents the most general formula for the [3/3] potential within the E-(a_1 + e) coupling scheme. It contains 5 force constants and 6 vibronic constants combined into A_1 and A_2 terms. With various approximations for A_1 and A_2, the simplified formulas (52), (55), (57), (58), (59) and (61) may be obtained. Notice that A_1 is constant in first-order, a function $A_1 = f(Q_1)$ in second-order and a function $A_1 = f(Q_1, \varrho)$ in third-order. Similarly, A_2 vanishes in first-order, it is a constant in second-order and a function $A_2 = f(Q_1)$ in third-order. If the admixture of the a_1 mode is switched off, within the E-e coupling scheme, A_1 is constant in the first and second order and a function $A_1 = f(\varrho)$ in the third order, as presented in Table 5.

Table 5. Various analytic formulas for APS

Coupling	Potential	A_1	A_2	Formula
E_g-e_g	[2/1]	const.	0	–
	[3/1]	const.	0	(57)
	[2/2]	const.	const.	(55)
	[3/2]	const.	const.	(58)
	[3/3]	$f(\varrho)$	const.	(61)
E_g-$(a_{1g} + e_g)$	[2/1]	const.	0	–
	[3/1]	const.	0	–
	[2/2]	$f(Q_1)$	const.	(52)
	[3/2]	$f(Q_1)$	const.	(59)
	[3/3]	$f(Q_1, \varrho)$	$f(Q_1)$	(67)

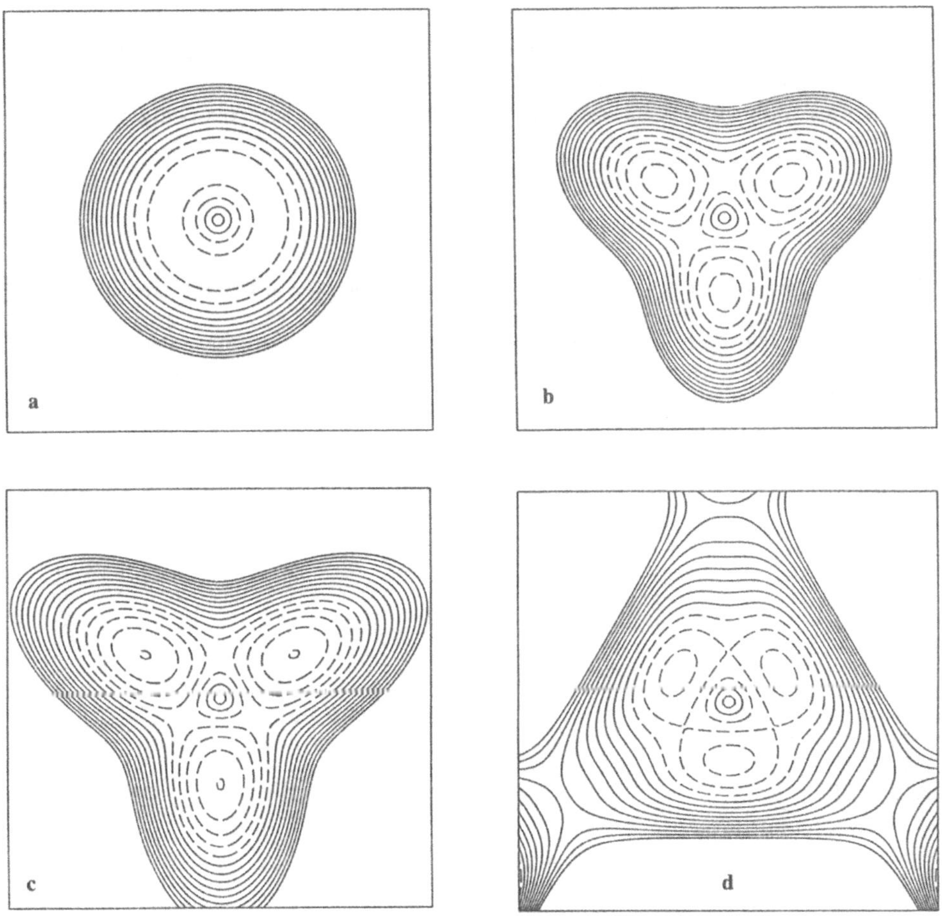

Fig. 6a–d. Computer plot of the adiabatic potential surface $E(Q_2, Q_3; [Q_1])$ in various degrees of complexity. $C_0 = 0.05$, $K_2 = 20.0$, $A = 3.0$. **a)** $B = 0$, $C = 0$, $K_3 = 0$; **b)** $B = 6.0$, $C = 0$, $K_3 = 0$; **c)** $B = 6.0$, $C = 10.0$, $K_3 = 0$; **d)** $B = 6.0$, $C = 10.0$, $K_3 = 40.0$

Restricting ourselves to Q_1 = const., the behaviour of the $E(Q_2, Q_3; [Q_1])$ potential is similar to the "Mexican hat" function with warping. Figure 6 shows a plot of the function

$$E = C_0 + K_2\varrho^2 + K_3U - [(A + C\varrho^2)^2\varrho^2 + B^2\varrho^4 + 2(A + C\varrho^2)BU]^{1/2} \quad (68)$$

with $U = 3\,Q_2^2Q_3 - Q_3^3 = -\varrho^3\cos(3\,\varphi)$ in various degrees of complexity. The following conclusions may be drawn from it:

(i) The linear vibronic constant A is responsible for a valley on the APS at the Jahn-Teller radius ϱ_0. Within the [2/1] approximation, $\varrho_0 = A/2\,K_2$ holds.

(ii) The quadratic vibronic constant B yields an additional warping with three minima and three saddle points. They are localized at $\varrho_0 = A/(2\,K_2 \pm 2\,B)$ and $\varphi_0 = n\pi/3$.

(iii) The cubic vibronic constant C magnifies the value of A so that the Jahn-Teller radius ϱ_0 increases. Simultaneously the slope of the APS is changed as seen from differences between isoenergy lines.

(iv) The cubic vibrational constant K_3 describes the anharmonicity effects. Notice, that both the quadratic and cubic approximations give an incorrect behaviour for larger displacements from the reference configuration; they are applicable only for small displacements as shown by Fig. 7.

Various types of vibronic coupling in various degrees of complexity are compared in Table 6.

Table 6. The dependence of number of APS constants on the type of vibronic interaction[a]

Vibronic interaction	Force constants linear	harmonic	cubic	Vibronic constants linear	quadratic	cubic
	$m = 1$	$m = 2$	$m = 3$	$n = 1$	$n = 2$	$n = 3$
Case 1						
T-e, T-t_2	0	1	1	1	1	1
T-(e + t_2)	0	2	3	2	3	4
T-(a_1 + e + t_2)	1	3	6	2	5	9
T-(a_1 + 2e + t_2)	1	5	12	3	9	19
T-(a_1 + e + 2t_2)	1	5	13	3	9	19
T-(a_1 + 2e + 2t_2)	1	7	21	4	14	34
Case 2						
E-b_1, E-b_2	0	1	0	1	0	1
E-(b_1 + b_2)	0	2	0	2	0	4
E-(a_1 + b_1 + b_2)	1	3	3	2	2	4
E-(a_1 + b_1 + 2b_2)	1	5	5	3	3	13
E-(a_1 + 2b_1 + b_2)	1	5	5	3	3	13
E-(a_1 + 2b_1 + 2b_2)	1	7	7	4	4	24
Case 3						
E-e	0	1	1	1	1	1
E-(a_1 + e)	1	2	3	1	2	3
E-(a_1 + 2e)	1	4	8	2	3	7

[a] The potential of [m/n] type results from combination of degrees of expansion in force (m) and vibronic (n) constants, respectively

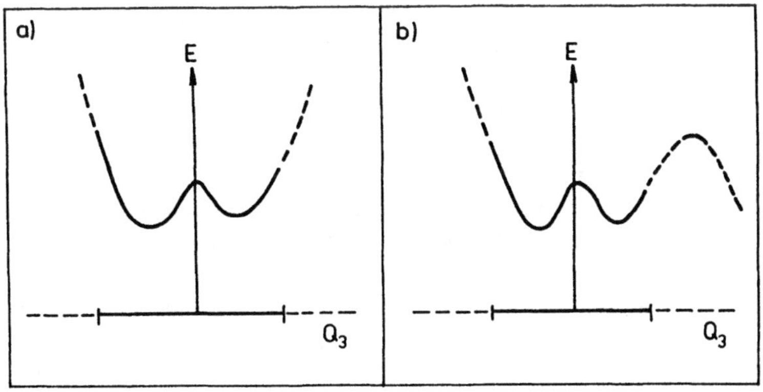

Fig. 7a, b. Cut of the adiabatic potential surface $E(Q_2, Q_3; [Q_1])$ for **a)** [2/2] potential; **b)** [3/3] potential

8 Symmetry Descent Concept

The determination of distortion routes for coordination polyhedra due to the Jahn-Teller effect represents the subject of numerous papers. The usual treatment is to investigate the analytic form of the adiabatic potential generated by the first-order perturbation theory. In terms of the above language, the eigenvalues of

$$U_{kl} = E_k^0 \delta_{kl} + H_{kl}^{(1)} + H_{kl}^{(2)} \qquad (69)$$

are analysed.

Some authors, on the other hand[32–34], applied the group theoretical approach to this problem; it is based on the symmetry properties of the Jahn-Teller active vibrational mode. Unfortunately, these studies have also been restricted to the first-order perturbation theory. This section exploits an entirely different treatment. The correlation among symmetry point groups is investigated for individual Jahn-Teller active systems from the viewpoint of the pertinent electronic state.

The Jahn-Teller theorem (see Sect. 4) implies the following consequences:

(i) With the exception of linear systems, the equilibrium geometry (characterized by the vanishing gradient components $K_i = 0$ and positive force constants $K_{ij} > 0$) corresponds to the nondegenerate electronic state S_N (thus being of A or B type).

(ii) The nondegenerate electronic state, stable with respect to the nuclear configuration (molecular geometry), arises by splitting the degenerate state S_D (being of E or T type, eventually of G or H type).

(iii) The splitting of the degenerate electronic state S_D is accompanied by a decrease of symmetry, i.e. some symmetry elements vanish.

In other words, the Jahn-Teller distortion causes a descent in the symmetry $G^0 \rightarrow G^n$ and the original electronic state $S_D(G^0)$ is split yielding the nondegenerate term $S_N(G^n)$. The symbol G^0 marks the symmetry point group of the reference system. A symmetry descent generates the n-th level subgroup G^n. For the first-level subgroup $G^1 \subset G^0$

holds, the second-level subgroup satisfies the relation $G^2 \subset G^1 \subset G^0$, etc. An important relation is valid among the group orders $\mathcal{O}$

$$\mathcal{O}(G^0) = m_1\mathcal{O}(G^1) = m_1 m_2 \mathcal{O}(G^2) = \ldots \tag{70}$$

The integer multiplicator m_i is equal to the number of equivalent configurations within the subgroup G^n to which the reference group G^{n-1} can distort. This number corresponds to the competent minimum points of the adiabatic potential surface.

These considerations may be summarized in the following statements:

(i) The symmetry elements of the perturbed geometry form a subgroup G^n of the reference system: $G^n \subset G^0$.

(ii) The irreducible representation $\Gamma(S^n)$ describing the electronic state in the "perturbed" geometry G^n originates in the splitting of the multidimensional irreducible representation, $\Gamma(S_D^0)$, which corresponds to the actual degenerate electronic state in the reference geometry G^0.

(iii) The stable (equilibrium) geometry corresponds to the nondegenerate electronic state, $S_N(G^n)$, described by the one-dimensional irreducible representation $\Gamma(S_N)$. Otherwise the system continues in symmetry descent.

These considerations are illustrated in Fig. 8. The results of the group-theoretical analysis of all molecular symmetry point groups have been summarized elsewhere[35, 36]. The procedure is illustrated for the symmetry group O_h in Fig. 9. This group contains 6 multidimensional irreducible representations (E_g, E_u, T_{1g}, T_{2g}, T_{1u}, T_{2u}) and has 24 subgroups. The distortion route depends on the actual electronic state. For example, the electronic term T_{2g} is split into B_{2g} and E_g terms of D_{4h}, or into A_{1g} and E_g terms of D_{3d} subgroup at the first level. The remaining first-level subgroups (O, T_d, T_h) cause no splitting of the T_{2g} term. At the second level the degenerate first-level terms (E or T) can be split to give nondegenerate states (the corresponding subgroups are given in parentheses): $B_2(D_4)$, $A_1(D_3)$, $B_2(D_{2d})$, B_{1g}, B_{2g} and $B_{3g}(D_{2h})$, $A_g(S_6)$, $A_1(C_{3v})$, A_g and $B_g(C_{2h})$. Analogously, at the third level the symmetry decrease yields: B_1, B_2 and $B_3(D_2)$, $A(C_3)$, B_1 and $B_2(C_{2v})$, $B_g(C_{2h})$, A and $B(C_2)$, $A_g(C_i)$, A′ and A″(C_s). At the fourth level only a single nondegenerate electronic term is possible: $B(C_2)$.

Restriction to the so-called Jahn-Teller active vibrations yields the distorted geometries D_{3d} and D_{4h} in accordance with earlier investigations[32–34]; the present approach is more general.

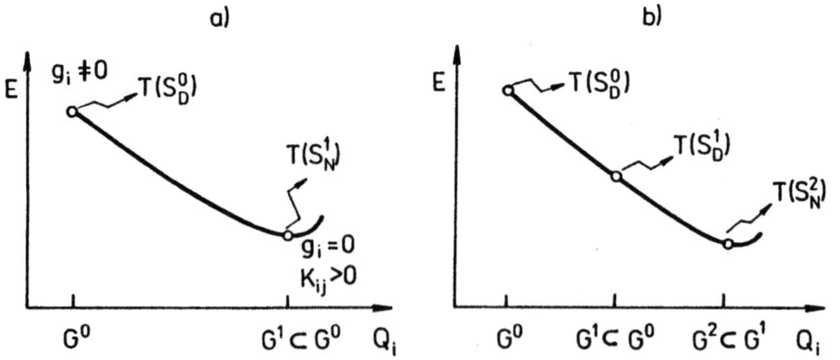

Fig. 8a, b. Possible routes of the Jahn-Teller distortions. **a** – single stage process, **b** – double stage process

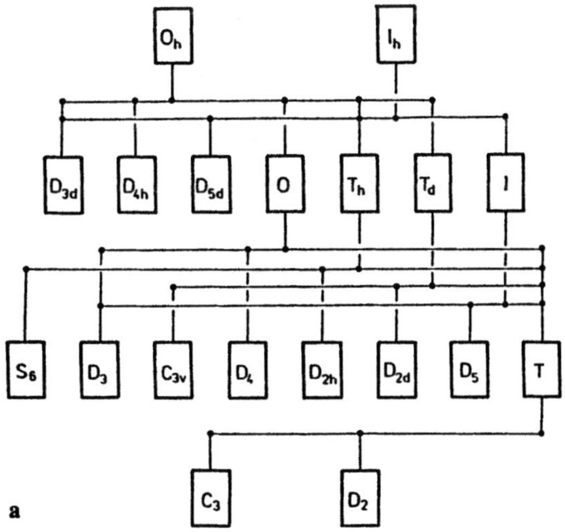

a

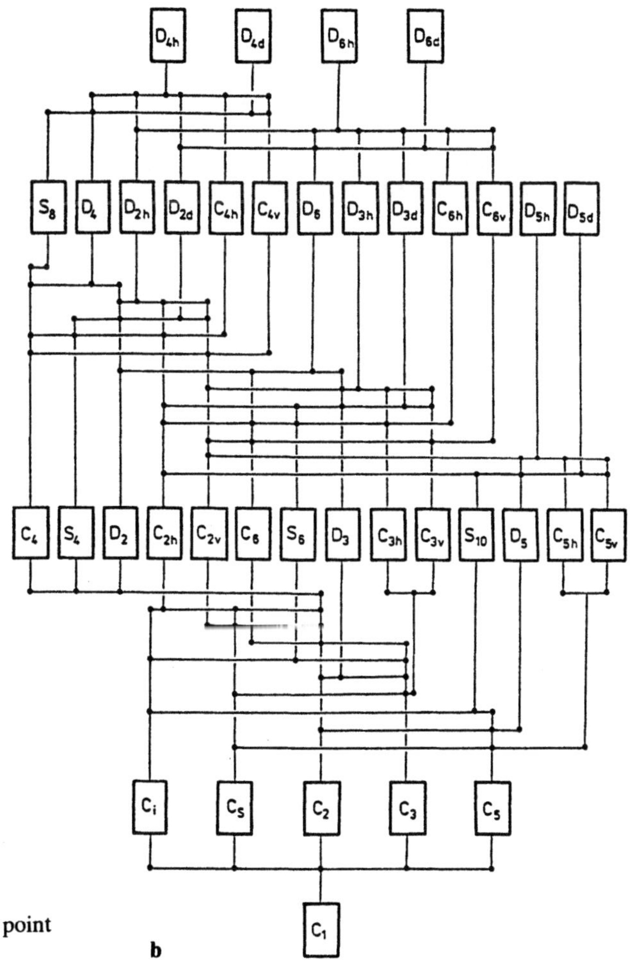

Fig. 9. Symmetry descent of the point groups to their subgroups

b

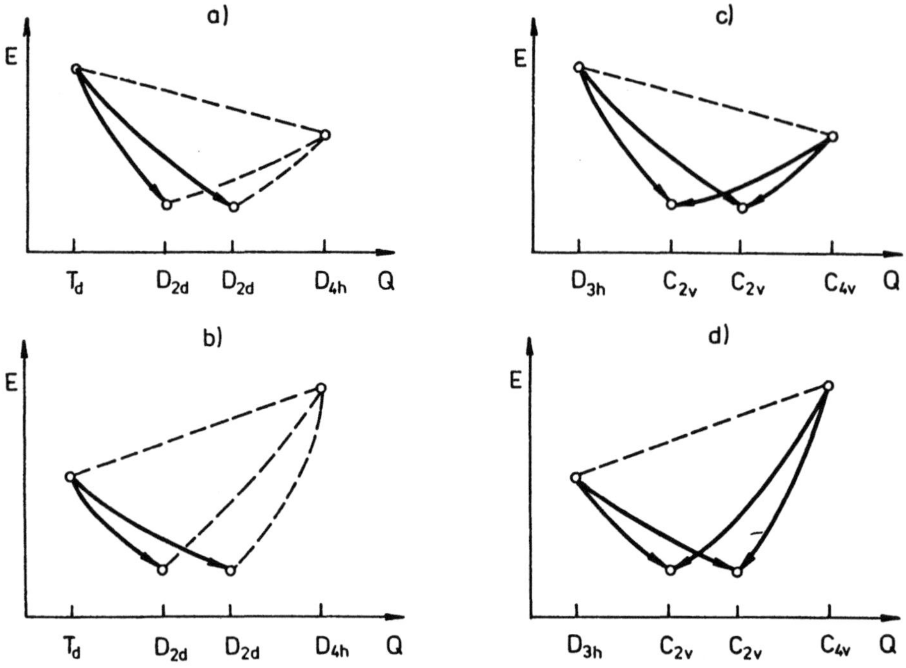

Fig. 10a–d. Forbidden --- and allowed —— Jahn-Teller distortions for $T_d \rightarrow D_{4h}$ (**a, b**) and $D_{3h} \rightarrow C_{4v}$ (**c, d**) geometric rearrangements

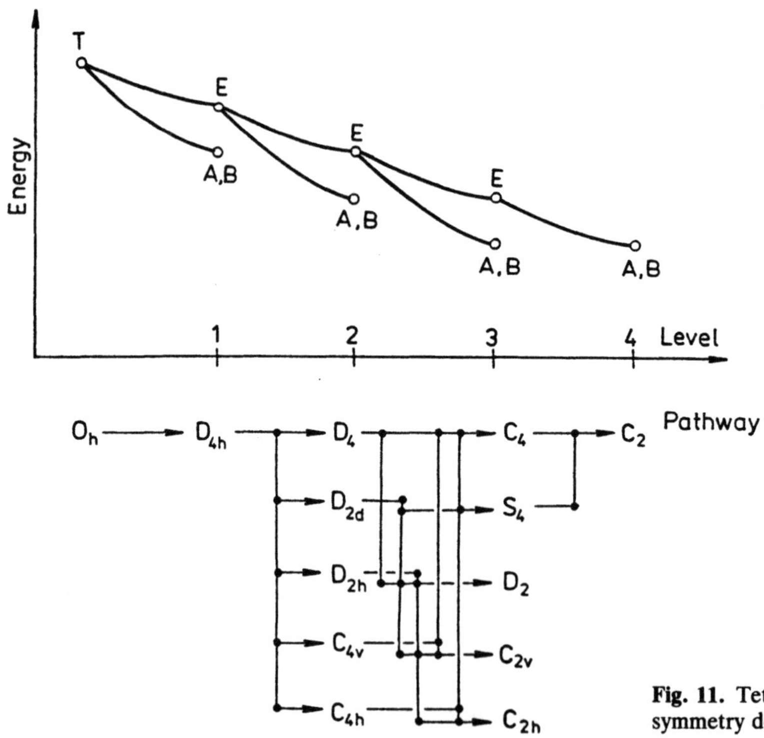

Fig. 11. Tetragonal route for symmetry descent of O_h group

On the basis of the above results, the geometrical rearrangement $T_d \rightarrow D_{4h}$ is not allowed by the Jahn-Teller effect since D_{4h} is not a correlating subgroup of the tetrahedron. The order of the D_{4h} group equals to 16 and the order of T_d is 24, so that condition (70) is not fulfilled. Analogously, the transition $D_{3h} \rightarrow C_{4v}$ should be excluded from the Jahn-Teller mechanism since $C_{4v} \not\subset D_{3h}$ (see Fig. 10).

For the symmetry-descent concept, an important question becomes relevant: Which molecular geometry should be considered as reference for operating the Jahn-Teller effect? For example, the octahedral geometry with T-type electronic state can split into O, T_h, T_d, D_{3d} or D_{4h} first-level subgroups. If the corresponding electronic state still remains degenerate, the symmetry descent continues to the second-, third- or fourth-level subgroups. One possible pathway of distortion is shown in Fig. 11 for the descent from cubic symmetry via the tetragonal pattern. The central-atom effect (the proton number influence of the central atom, as well as the oxidation and spin state of the complex) and the ligand-field effect represent important factors that determine the final ground-state geometry of the system. The vibronic coupling in hexacoordinate systems may refer to T_{2g}-(a_{1g} + e_g + t_{2g}) type within the group O_h, to E-(a_1 + b_1 + b_2) type within the group D_4, to E-(a_1 + e) type within the group C_3, etc. Many possibilities come into consideration.

Liehr[37)] postulated that the vibronic coupling operates in the presence of electronic degeneracy within the group of the highest symmetry (maximum group order). Nevertheless, no proof of this theorem has been presented. On the contrary, quantum-chemical calculations showed[25)] that the vibronic coupling should be considered within the reference group of the minimum symmetry (minimum group order) where the electronic state is still degenerate. For example, a better description of the adiabatic potential is obtained for certain MX_6^q systems if either E_g-(a_{1g} + e_g) coupling within the group D_{3d} or E_g-(a_1 + b_1 + b_2) coupling within the group D_{4h} are considered instead of T_{2g}-(a_{1g} + e_g) within the group O_h.

9 Experimental Structure Data

Several extensive reviews are available[2, 11, 12, 16)] in which various consequences of the vibronic interactions are discussed. Therefore only a few examples of the structural consequences of vibronic interactions in metal complexes are given below.

Hexachlorocomplexes serve as an example of the static Jahn-Teller effect. The metal-ligand distances within the coordination polyhedron usually obey a relation of

$$R_e(2\times) \approx R_e'(2\times) \ll R_a(2\times) \quad (71)$$

so that two axial distances are elongated relative to the equatorial ones. This situation originates in the E_g-e_g, or more precisely in the E_g-(a_{1g} + e_g) vibronic coupling for the reference group $G^0 = O_h$. The tetragonality parameter $T = R_e/R_a$ for $CuCl_6^{4-}$ systems spans the interval of 0.75 to 0.87 and the assymetry parameter $\Delta = R_a - R_e$ ranges from 0.36 to 0.77 (10^{-10} m). The corresponding structure data have been collected in Ref. 40. Although the observed tetragonal distortions agree with the Jahn-Teller theorem, it was

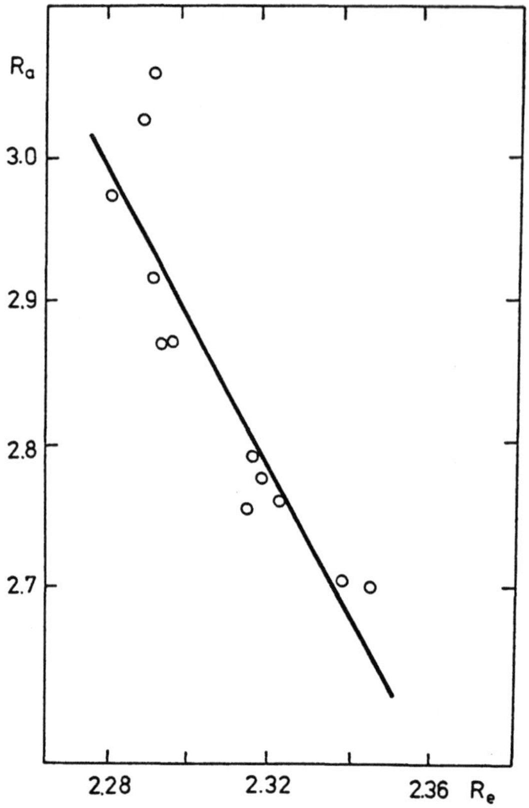

Fig. 12. R_a vs. R_e dependence for hexachlorocupric complexes (in 10^{-10} m)

not understood why the axial and mean equatorial metal-ligand distances were correlated along a smooth curve – see Fig. 12 (the R_a versus R_e dependence) – until the concept of equatorial-axial interactions was formulated[8,9].

Similar collections have been published for hexafluorocomplexes[38], hexanitrocomplexes[39] and hexaaquacomplexes[40].

If only the tetragonal distortion is considered for the [2/2] type adiabatic potential, $Q_2 = 0$ and $\cos(3\varphi) = \pm 1$ are valid in Eq. (52). The corresponding formula relaxes to

$$E = E^0 + \tfrac{1}{2} K_{aa} Q_1^2 + \tfrac{1}{2} K_{ee} Q_3^2 \pm [(A + ZQ_1)Q_3 + BQ_3^2] \tag{72}$$

A behaviour of such a paraboloid potential is better visualized in the basis set of internal coordinates R_a and R_e where a harmonic form of

$$E = E^0 + bR_a + cR_a^2 + dR_e + eR_aR_e + fR_e^2 \tag{73}$$

is obtained. This form leads to the following conditions for the local energy minima

$$(\partial E/\partial R_a)_{R_e} = b + eR_e + 2\,cR_a = 0 \tag{74a}$$

$$(\partial E/\partial R_e)_{R_a} = d + eR_a + 2\,fR_e = 0 \tag{74b}$$

so that an R_a vs R_e dependence adopts a form of two descending straight lines

$$R_a = -(e/2c)R_e - b/2c \qquad \text{for } R_a > R_e \qquad (75a)$$

$$R_a = -(2f/e)R_e - d/e \qquad \text{for } R_a < R_e \qquad (75b)$$

This yields an unacceptable prediction of $R_e \leq 0$ for larger displacements of R_a. Therefore, at least the cubic or higher-order terms must be considered (e.g. [3/2] potential) to obtain a correct (hyperbolic) form of the R_a vs R_e dependence.

The theory of step-by-step descent in symmetry due to the Jahn-Teller effect may be applied to real solid-state systems where the symmetry of the unit cell is determined by the symmetry of the Jahn-Teller active centre. Phase transitions in such cases where the Jahn-Teller mechanism is determining must correspond to the descent in symmetry via one or more chains originating at the same parent symmetry group G^0.

This theory is applicable also to pseudo-Jahn-Teller systems if we understand the pseudo-Jahn-Teller effect as the Jahn-Teller effect in excited electronic states.

Let us show the application of this theory by a few examples.

$K_2PbCu(NO_2)_6$. There are three structural phases of this compound. The high-temperature α phase is of $Fm3(T_h^3)$ symmetry (structure determined at 300 K). On lowering the temperature, the incommensurate β phase of $Fmmm(D_{2h}^{23})$ symmetry is obtained (structure determined at 275 K). This chain, $T_h \rightarrow D_{2h}$, corresponds to the splitting of doubly degenerate electronic term. The low-temperature γ phase of $C\bar{1}(C_i^1)$ symmetry (structure determined at 250 K) corresponds to another chain of splitting: $T_h \rightarrow S_6 \rightarrow C_i$ because of an unstable S_6 group with a doubly degenerate electronic term. Thus the existence of these phases may be explained as a consequence of a Jahn-Teller descent in symmetry of the parent (high-temperature) phase in two different chains. Figure 13 illustrates such a behaviour. In fact, the α phase exhibits a dynamic Jahn-Teller effect, the β phase a planar dynamic effect, and the γ phase a static Jahn-Teller effect.

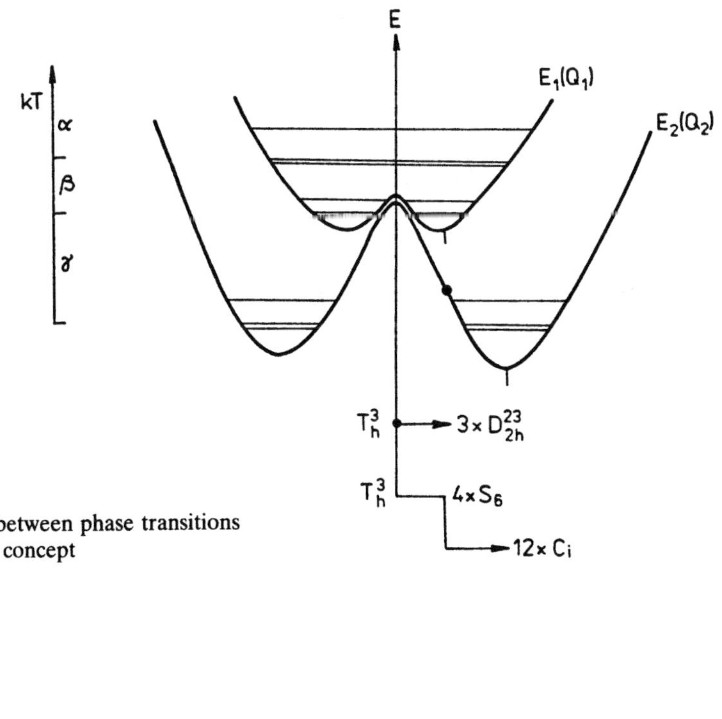

Fig. 13. Relationship between phase transitions and symmetry descent concept

$BaTiO_3$. This compound is known as a pseudo-Jahn-Teller system so we must account for its first excited electronic state $^1T_{1u}$. The high-temperature phase[42] (over 393 K) of this compound has Pm3m(O_h^1) symmetry. The following ferroelectric phases are known at lower temperatures[42]: P4mm(C_{4v}^1) under 393 K, Cmm2(C_{2v}^{14}) under 278 K and R3m(C_{3v}^5) under 183 K. Their existence may be explained due to Jahn-Teller splitting of the triple degenerate electronic state in two chains:
(i) $O_h \rightarrow D_{4h} \rightarrow C_{4v} \rightarrow C_{2v}$ (unstable D_{4h} and C_{4v} configurations with double electronic degeneracy)
(ii) $O_h \rightarrow T_d \rightarrow C_{3v}$ (unstable T_d configuration with triple electronic degeneracy)
The existence of the P4mm(C_{4v}^1) phase may be understood as a consequence of dynamic effects: superposition of two C_{2v} configurations. Another high-temperature phase of P6$_3$/mmc(D_{6h}^4) symmetry over 1603 K may be explained as a superposition of C_{3v} and C_{2v} symmetries. The alternative treatment explains the situation in another two chains:
(i) $D_{6h} \rightarrow D_{3h} \rightarrow C_{2v}$ (unstable D_{3h} symmetry with double electronic degeneracy)
(ii) $D_{6h} \rightarrow D_{3h} \rightarrow C_{3v}$ (the same holds for D_{3h} symmetry).
Here the existence of other phases (C_{4v}, O_h) is explained as a consequence of dynamic effects. Thus the existence of all the mentioned phases results partly from the Jahn-Teller descent in symmetry in two chains and partly from dynamic effects.

$NaNbO_3$. There are 7 structural phases of this compound[42]: Pm3m(O_h^1) over 913 K, P4/mbm(D_{4h}^5) over 848 K, Cmmc(D_{2h}^{17}) over 793 K, two Pmm(D_{2h}^{13}) phases over 753 K and over 643 K, Pbma(D_{2h}^{11}) at room temperature and finally R3c(C_{3v}^6) phase under 173 K. Their existence may be understood due to the Jahn-Teller descent in symmetry via the following chains (the triply degenerate electronic state is split):
(i) $O_h \rightarrow D_{4h}$
(ii) $O_h \rightarrow D_{4h} \rightarrow D_{2h}$ (unstable D_{4h} symmetry with double electronic degeneracy) with two possible D_{2h} configurations
(iii) $O_h \rightarrow T_h \rightarrow D_{2h}$ (unstable T_h symmetry with triple electronic degeneracy)
(iv) $O_h \rightarrow T_d \rightarrow C_{3v}$ (unstable T_d symmetry with triple electronic degeneracy)
The first case may be alternatively explained as a dynamic superposition of two D_{2h} symmetries. The second and third cases are connected with the existence of different D_{2h} phases.

$KNbO_3$. One could expect the similarity of this compound and $NaNbO_3$. In reality, there are 4 phases[42]: Pm3m(O_h^1) over 683 K, P4mm(C_{4v}^1) over 483 K, Bmm2(C_{2v}^{14}) over 263 K and R3m(C_{3v}^5). Thus the situation is similar to the case of $BaTiO_3$.

10 Molecular Orbital Calculations of Jahn-Teller Coupling Constants

There are two methods of evaluating the Jahn-Teller coupling constants. The former is based on their direct evaluation on various levels of complexity, namely
(i) the electrostatic model of van Vleck[43],
(ii) crystal field approach[44, 45],
(iii) angular overlap model (AOM)[15, 46–48],

(iv) Nikifirov's analysis of LCAO contributions[49, 50],
(v) analytical derivatives within LCAO approach[51],
The latter method is based on a least-squares fitting of the numerical map of the adiabatic potential to the analytic form[24–26, 52, 53] and is applicable in an arbitrary degree of complexity.

The calculation of any Jahn-Teller coupling constant can be carried out via the evaluation of matrix elements

$$c_\gamma^\Gamma = \left\langle \phi_{\gamma_i}^{\Gamma_i} \left| \left(\frac{\partial \hat{V}}{\partial Q_\gamma^\Gamma} \right)_0 \right| \phi_{\gamma_j}^{\Gamma_j} \right\rangle \tag{76}$$

having the appropriate symmetry-adapted wave functions $\phi_{\gamma_i}^{\Gamma_i}$, $\phi_{\gamma_j}^{\Gamma_j}$; Q_γ^Γ are symmetry coordinates corresponding to the γ component of the Γ irreducible representation and

$$\frac{\partial \hat{V}}{\partial Q_\gamma^\Gamma} = \sum_{n=1}^{N} \frac{\partial \hat{V}}{\partial r_n} \frac{\partial r_n}{\partial Q_\gamma^\Gamma} \tag{77}$$

where r_n stands for the generic coordinate (R_n, ϑ_n, φ_n) of the n-th ligand.

Within the approximation of crystal-field theory, the Jahn-Teller coupling constants may be evaluated in a simple way. For example, in the case of E-e vibronic coupling in octahedral MX_6 complexes[44, 45]

$$A = \left(\frac{\partial V_{22}}{\partial Q_3} \right)_0 = -\frac{2\sqrt{3}}{\sqrt{7}} q \left[\frac{\partial F_2(R)}{\partial R} + \frac{5}{12} \frac{\partial F_4(R)}{\partial R} \right]_{R=R_0} \tag{78}$$

with

$$F_k(R) = R^{-(k+1)} \int_0^R r^k f_{nl}^2(r) r^2 dr + R^k \int_R^\infty r^{-(k+1)} f_{nl}^2(r) r^2 dr \tag{79}$$

where $f_{nl}(r)$ is the radial part of the electronic function with n and l quantum numbers; R_0 is the reference M-X distance; q is the effective ligand charge.

More sophisticated results were obtained by the angular-overlap method (AOM)[15, 46–48]. The formula (76) may be rewritten in the form

$$c_\gamma^\Gamma = \sum_{\lambda,\omega} \sum_{n=1}^{N} \frac{\partial}{\partial r_n} [e_{\lambda n} F_{\lambda\omega}(\phi_i, X_n)\ F_{\lambda\omega}(\phi_j, X_n)] \frac{\partial r_n}{\partial Q_\gamma^\Gamma} \tag{80}$$

where $F_{\lambda\omega}(\phi_i, X_n)$ is the angular-overlap factor at a given M-X_n distance; $e_{\lambda n}$ is the radial parameter; λ indicates the bonding symmetry (σ, π, δ); ω specifies the particular orbital and r_n is defined in Eq. (77).

It follows from Eq. (80) that the Jahn-Teller coupling constants related to the stretching vibrational modes are functions of $\partial e_{\lambda n}/\partial R$ while those related to the bending modes are simply functions of $e_{\lambda n}$. $F_{\lambda\omega}$ depends upon the bond length R and angular coordinates ϑ and φ, respectively.

The quadratic Jahn-Teller coupling constants can be easily obtained by introducing second derivatives into (76) and (80). The results of d^x systems for a number of symmetries were published by Bacci[15, 46, 47] and Warren[48]. The vibronic coupling constants can be expressed through a limited number of parameters in the framework of AOM. It is possible to calculate e_λ and its derivatives theoretically, but better results are obtained with e_λ derived from experiments. Their derivatives can be evaluated numerically. The coupling constants related to the stretching modes are less accurate than those related to the bending modes, because of the uncertainity about their derivatives. Calculation of force constants remains problematic since only approximate values can be reached for less-simple systems.

Nikiforov and coworkers[49, 50] tried to evaluate Jahn-Teller coupling constants of 3 d transition metals in the cluster approximation by using a semiempirical MO LCAO approach.

A qualitatively new treatment was proposed by Dixon[52], who fitted the electronic energies to polynomial functions of displacement coordinates. Pelikán and coworkers[24–26, 53] independently elaborated the method of evaluation of Jahn-Teller coupling and force constants using a non-linear least-squares fitting of total energy of the system. This treatment consists of two steps. First, the numerical map of APS values should be calculated by any quantum-chemical method. Then, the least-squares method is applied in order to fit the numerical map, $E^c(Q_i)$, to the analytic form. Actually, the set of approximate values, $E^a(Q_i)$, is produced so that the functional

$$\mathcal{F} = \sum_i w_i^2[E^c(Q_i) - E^a(Q_i)]^2 \tag{81}$$

adopts a minimum value. The values of $E^a(Q_i)$ are calculated by Eq. (44) for a trial set of potential (i.e. coupling and force) constants. The statistical weights were chosen in accordance with the metric weighting concept

$$w_i = \left[\sum_k Q_k^2\right]^{-1/2} \tag{82}$$

where Q_k is a set of displacement coordinates for the given APS point. The established minimization procedure, in general, is based on a non-linear optimization (Newton-Raphson and/or Fletcher-Powell algorithms were used). In order to accelerate a convergence, the process may be split into two, linear and non-linear optimizations. Another simplification of the problem consists in considering only selected cuts of the APS (e.g. only D_{4h} distortions of ML_6 systems). This may lead to a simpler, purely polynomial form of the APS that can be fitted by the linear optimization only[26]. The quality of regression is measured by statistical characteristics (e.g. standard deviations of individual potential constants, the correlation coefficient, the discrepancy R-factor).

The above method was systematically applied to APS of some hexahalo complexes of the first transition-metal row. The total energy of the system has been obtained by CNDO-UHF calculations. Selected results are presented in Table 7.

A large number of results[24–26] for MX_6^q (M = Cr, Mn, Fe, Co, Ni, Cu; X = F, Cl, Br; q = − 3, − 4) systems with double or triple electron degeneracy allows us to conclude:

(i) Inclusion of the total symmetric vibrational mode manifests itself in a significant contribution to the values of the corresponding vibronic constants. Thus, its importance is clearly demonstrated.
(ii) The e_g vibrational modes are much softer than the a_{1g} ones according to the calculated force constants: $K_{ee} < K_{aa}$. This fact may play an important role in deformations of the solid-state systems.

Table 7. The calculated values of potential constants[a] and characteristics[b] of APS extreme points[24)]

System	$^5MnF_6^{3-}$	$^5MnCl_6^{3-}$	$^5MnBr_6^{3-}$
A	-1.565	-0.9008	-1.005
B	-0.65	-0.064	0.11
Z	-2.387	-1.219	-1.56
K_{aa}	28.676	25.5600	25.5599
K_{ee}	20.370	17.4198	18.953
T_{aaa}	-6.900	-5.290	-5.533
T_{aee}	-23.323	-20.95	-30.31
T_{eee}	-4.80	-3.036	-1.69
Correl. coeff.	0.999991	0.999999	0.999999
R-factor	0.00276	0.00113	0.00154
Q_1^0	2.10266	2.40786	2.55364
Minimum:			
ΔQ_1	-0.0010	-0.0004	0.00004
ϱ	0.077	0.052	0.055
φ	0	π	π
ΔE	-0.0618	-0.02355	-0.0273
Saddle point:			
ΔQ_1	-0.0021	-0.00017	-0.0001
ϱ	0.076	0.0507	0.052
φ	π	0	0
ΔE	-0.0585	-0.02305	-0.0261

[a] Units used: A is in 10^{10} eVm^{-1}; B, Z, K_{aa} and K_{ee} are in 10^{20} eVm^{-2}; T_{aaa}, T_{aee} and T_{eee} are in 10^{30} eVm^{-3}. The R-factor is defined as $R = [\sum_i (E_i^a - E_i^c)^2 / \sum_i (E_i^a)^2]^{1/2}$

[b] $\Delta Q_1 = Q_1 - Q_1^0$, $\varrho = (Q_2^2 + Q_3^2)^{1/2}$, $\varphi = \mathrm{arctg}\,(Q_2/Q_3)$, $\Delta E = E^{ext}(Q) - E(Q^0)$. Distances are in 10^{-10} m; energies in eV

(iii) The Jahn-Teller distortions calculated from these constants are small when compared with distortions found for solid-state complexes. It indicates that solid-state influences may prevail over the net Jahn-Teller effect.

(iv) Calculated tetragonal distortions and stabilization energies in systems with triple electron degeneracy are much lower in comparison with analogous complexes with double electron degeneracy.

(v) Higher distortions and stabilization energies occur in M(III) complexes compared to analogous M(II) systems. All the characteristics calculated are also functions of the polarity of metal-ligand bonds, spin multiplicity and proton number of the central atom.

Despite of the simple CNDO/2 method used the trends in the calculated potential constants, distortions and stabilization energies seem correct. The solid state, however, contributes to these quantities as well, and it probably amplifies the degree of distortion.

11 Conclusions

The Jahn-Teller effect represents an important concept for investigating the structure and characteristics of materials. This phenomenon has been studied by many authors with various methods, and it has been the subject of many monographs.

The main purpose of this review is to draw attention to some theoretical aspects that have been very simplified in the interpretation of experimental results. Among them, the derivation of the analytical formula describing the shape of the APS is of great importance. Therefore, side by side with some details of the perturbation theory, we tried to explain the method of deriving the higher-order terms of the vibronic expansion and the interrelation of various types of coupling coefficients. The importance of individual expansion terms is also clearly demonstrated. The problem of reference geometry is analysed in connection with the theory of step-by-step descent in symmetry. This theory is able to predict possible symmetry changes caused by the Jahn-Teller effect.

We concentrated on the stereochemical consequences of the Jahn-Teller effect (phase transitions, R_a vs. R_e dependence in hexacoordinate complexes), but the ideas of our studies are applicable to other experimental aspects of the Jahn-Teller effect.

Last, but not least, is the problem of evaluation of Jahn-Teller coupling constants. It is evident that the use of more sophisticated techniques – in order to obtain more reliable results via first-principles calculations – implies a higher complexity of the resulting expressions. On the other hand, the method of least-squares fitting needs much more computer time; but it is universal (suitable for all quantum-chemical methods), and the reliability of the results is given only by the reliability of the quantum-chemical method used (and by the complexity of the set of the potential constants used).

Finally, we must warn against generalization of the concept of the Jahn-Teller effect. Many experimental results have been "explained" in terms of this effect, but some of them may be interpreted in an alternative way. We must analyse very carefully what can and what cannot be a consequence of Jahn-Teller effect.

12 Appendix. Reduction of Matrix Elements

This Appendix is concerned with an appropriate expression for the vibronic matrix elements V_{kl}^{r} (41) and V_{kl}^{rs} in the basis set of degenerate electronic wave functions $|k\rangle$ and $|l\rangle$. The operator parts

$$\frac{\partial V_{eN}}{\partial Q_r} = \dot{V}\begin{pmatrix}\Gamma_r\\ \gamma_r\end{pmatrix} \tag{A1}$$

$$\frac{\partial^2 V_{eN}}{\partial Q_r \partial Q_s} = \ddot{V}\begin{pmatrix}\Gamma_r & \Gamma_s\\ \gamma_r & \gamma_s\end{pmatrix} \tag{A2}$$

are classified according to the irreducible representations, Γ_r, and their components, γ_r, within the symmetry group G_0. A similar classification is used for wave functions $\left|\begin{matrix}\Gamma_k\\ \gamma_k\end{matrix}\right\rangle$ and $\left|\begin{matrix}\Gamma_l\\ \gamma_l\end{matrix}\right\rangle$.

Thus, the Wigner-Eckart theorem states

$$V_{kl}^{r} = \left\langle \begin{matrix}\Gamma_k\\ \gamma_k\end{matrix} \middle| \dot{V}\begin{pmatrix}\Gamma_r\\ \gamma_r\end{pmatrix} \middle| \begin{matrix}\Gamma_l\\ \gamma_l\end{matrix} \right\rangle = N \left\langle \begin{matrix}\Gamma_l & \Gamma_r\\ \gamma_l & \gamma_r\end{matrix} \middle| \begin{matrix}\Gamma_k\\ \gamma_k\end{matrix} \right\rangle \cdot X(\Gamma_k\Gamma_l\,;\Gamma_r) \tag{A3}$$

The reduced matrix element $X(\Gamma_k\Gamma_l\,;\Gamma_r) = X_{\Gamma_r}$ depends only on Γ; it does not depend on the γ components of the multidimensional representations. The Clebsh-Gordan coefficients $\left\langle \begin{matrix}\Gamma_k & \Gamma_l\\ \gamma_k & \gamma_l\end{matrix} \middle| \begin{matrix}\Gamma_r\\ \gamma_r\end{matrix} \right\rangle$ (CG) couple the angular momenta of $\left| \begin{matrix}\Gamma_k\\ \gamma_k\end{matrix} \right\rangle$ and $\left| \begin{matrix}\Gamma_l\\ \gamma_l\end{matrix} \right\rangle$ to yield the angular momentum of $\left| \begin{matrix}\Gamma_r\\ \gamma_r\end{matrix} \right\rangle$, so that

$$\left| \begin{matrix}\Gamma_r\\ \gamma_r\end{matrix} \right\rangle = \sum_{\gamma_k,\gamma_l} \left\langle \begin{matrix}\Gamma_k & \Gamma_l\\ \gamma_k & \gamma_l\end{matrix} \middle| \begin{matrix}\Gamma_r\\ \gamma_r\end{matrix} \right\rangle \left| \begin{matrix}\Gamma_k\\ \gamma_k\end{matrix} \right\rangle \left| \begin{matrix}\Gamma_l\\ \gamma_l\end{matrix} \right\rangle \tag{A4}$$

holds.

The Clebsh-Gordan coefficients are proportional to the 3 j-symbols

$$\begin{pmatrix} j_1 & j_2 & j_3\\ m_1 & m_2 & m_3\end{pmatrix} = (-1)^{j_1-j_2-m_3}\,(2\,j_3+1)^{-1/2} \left\langle \begin{matrix} j_1 & j_2\\ m_1 & m_2\end{matrix} \middle| \begin{matrix} j_3\\ -m_3\end{matrix}\right\rangle =$$

$$= (-1)^{j_1-j_2-m_3}\,[(j_1+j_2-j_3)!(j_1-j_2+j_3)!(-j_1+j_2+j_3)!]^{1/2} \times$$

$$\times\,[(j_1+m_1)!(j_1-m_1)!(j_2+m_2)!(j_2-m_2)!(j_3+m_3)!(j_3-m_3)!]^{1/2} \times$$

$$\times\,[(j_1+j_2+j_3+1)!]^{-1/2} \sum_k (-1)^k/[k!(j_1+j_2-j_3-k)! \times$$

$$\times\,(j_1-m_1-k)!(j_3-j_2+m_1+k)!(j_3-j_1-m_2+k)!] \tag{A5}$$

which possess important symmetry properties and are easily evaluable.

As a consequence of (A3), the replacement theorem is valid: if two operators, say $\dot{V}$ and $\dot{W}$, are of the same symmetry (i.e. they are of the same type with respect to an angular momentum), then both are reduced by the same Clebsh-Gordan coefficient C

$$\begin{aligned} V_{kl}^{r} &= C \cdot X_r\\ W_{kl}^{r} &= C \cdot Y_r \end{aligned} \tag{A6}$$

This implies a proportionality of

$$V_{kl}^{r} = W_{kl}^{r}\,\frac{X_r}{Y_r} \tag{A7}$$

(the method of operator equivalents). An angular-momentum operator may be chosen in the role of W_{kl}^{r}. Thus the transformation properties of spherical harmonics $Y_{j,m}(\vartheta, \varphi)$ are sufficient in evaluating the integrals (A6). Table A1 shows real combinations Y^R of

Table A1. Real linear combinations, Y^R, of spherical harmonics, $Y_{j,m}$

j	Y^R	Symmetry properties
0	$Y_{0,0}$	$x^2 + y^2 + z^2$
1	$\frac{-1}{i\sqrt{2}}(Y_{1,1} + Y_{1,-1})$	y
	$\frac{-1}{\sqrt{2}}(Y_{1,1} - Y_{1,-1})$	x
	$Y_{1,0}$	z
2	$Y_{2,0}$	z^2
	$\frac{-1}{i\sqrt{2}}(Y_{2,1} + Y_{2,-1})$	yz
	$\frac{-1}{\sqrt{2}}(Y_{2,1} - Y_{2,-1})$	xz
	$\frac{1}{\sqrt{2}}(Y_{2,2} + Y_{2,-2})$	$x^2 - y^2$
	$\frac{1}{i\sqrt{2}}(Y_{2,2} - Y_{2,-2})$	xy

spherical harmonics that enable the symmetry classification in terms of cartesian components. Since they are listed in all character tables, the classification of $\left|\begin{matrix}\Gamma\\ \gamma\end{matrix}\right\rangle$ in terms of angular momenta quantum numbers (j, m) is straightforward[1].

In order to illustrate the above consideration, let us evaluate the coupling coefficients for the doubly degenerate E term within the O group. The real basis functions are

$$z^2 \quad \ldots\ldots \left|\begin{matrix}E\\ \vartheta\end{matrix}\right\rangle = Y_{2,0} \tag{A7a}$$

$$x^2 - y^2 \ldots\ldots \left|\begin{matrix}E\\ \varepsilon\end{matrix}\right\rangle = (Y_{2,2} + Y_{2,-2})/\sqrt{2} \tag{A7b}$$

The only non-zero 3 j-symbols are $\begin{pmatrix}2 & 2 & 2\\ m_1 & m_2 & m_3\end{pmatrix}$ with $m_1 = m_2 + m_3$. For $m_i = 0, 2$ or -2 the coefficients adopt a value of $a = (2/35)^{1/2}$ except $\begin{pmatrix}2 & 2 & 2\\ 0 & 0 & 0\end{pmatrix} = -(2/35)^{1/2} = -a$. Therefore

$$\left\langle\begin{matrix}E & E\\ \vartheta & \vartheta\end{matrix}\middle|\begin{matrix}E\\ \vartheta\end{matrix}\right\rangle = (-1)^2\begin{pmatrix}2 & 2 & 2\\ 0 & 0 & 0\end{pmatrix} = -a \tag{A8}$$

and similarly

$$\left\langle\begin{matrix}E & E\\ \varepsilon & \varepsilon\end{matrix}\middle|\begin{matrix}E\\ \vartheta\end{matrix}\right\rangle = \left\langle\begin{matrix}E & E\\ \vartheta & \varepsilon\end{matrix}\middle|\begin{matrix}E\\ \varepsilon\end{matrix}\right\rangle = \left\langle\begin{matrix}E & E\\ \varepsilon & \vartheta\end{matrix}\middle|\begin{matrix}E\\ \vartheta\end{matrix}\right\rangle = a \tag{A9}$$

1 In some cases two or more sets of cartesian components (say $x^2 - y^2$, xy or xz, yz in D_3) belong to the same multidimensional irreducible representation so that a linear combination of them should be considered. Hence, the symmetry descent technique is applicable.

The normalized coupling coefficients are $c_i = N\, a_i$ where the normalization factor is

$$N = \left[\sum_i (a_i)^2 \right]^{-1/2} \tag{A10}$$

and i runs over the single component γ or over all components of the irreducible representation. Conventionally, $c = 1/\sqrt{2}$ for the former case.

Because of the renormalization, the same result is obtained using the Racah V-coefficients, Clebsh-Gordan coefficients or Wigner 3 j-symbols. The coupling coefficients have been tabulated in specialized monographs[54–57].

It must be mentioned that the same coupling coefficients are obtained for isomorphous groups (e.g. for O and T_d, C_{nv} and D_n, etc.). Moreover, the coupling coefficients for direct product groups ($T_h = C_i \otimes T$, $O_h = C_i \otimes O$, etc.) can be expressed by multiplying the coupling coefficients of the respective subgroups.

For the doubly degenerate electronic state ($\Gamma_k = \Gamma_l = E$), the relevant matrix elements for the cubic groups become

$$\begin{pmatrix} \left\langle \begin{matrix} E \\ \vartheta \end{matrix} \middle| \dot{V}\begin{pmatrix} e \\ \vartheta \end{pmatrix} \middle| \begin{matrix} E \\ \vartheta \end{matrix} \right\rangle ; \left\langle \begin{matrix} E \\ \vartheta \end{matrix} \middle| \dot{V}\begin{pmatrix} e \\ \vartheta \end{pmatrix} \middle| \begin{matrix} E \\ \varepsilon \end{matrix} \right\rangle \\ \left\langle \begin{matrix} E \\ \varepsilon \end{matrix} \middle| \dot{V}\begin{pmatrix} e \\ \vartheta \end{pmatrix} \middle| \begin{matrix} E \\ \vartheta \end{matrix} \right\rangle ; \left\langle \begin{matrix} E \\ \varepsilon \end{matrix} \middle| \dot{V}\begin{pmatrix} e \\ \vartheta \end{pmatrix} \middle| \begin{matrix} E \\ \varepsilon \end{matrix} \right\rangle \end{pmatrix} = \begin{pmatrix} -X_e/\sqrt{2}\ ;\ 0 \\ 0\ ;\ X_e/\sqrt{2} \end{pmatrix} = X_e C_3 \tag{A11}$$

$$\begin{pmatrix} \left\langle \begin{matrix} E \\ \vartheta \end{matrix} \middle| \dot{V}\begin{pmatrix} e \\ \varepsilon \end{pmatrix} \middle| \begin{matrix} E \\ \vartheta \end{matrix} \right\rangle ; \left\langle \begin{matrix} E \\ \vartheta \end{matrix} \middle| \dot{V}\begin{pmatrix} e \\ \varepsilon \end{pmatrix} \middle| \begin{matrix} E \\ \varepsilon \end{matrix} \right\rangle \\ \left\langle \begin{matrix} E \\ \varepsilon \end{matrix} \middle| \dot{V}\begin{pmatrix} e \\ \varepsilon \end{pmatrix} \middle| \begin{matrix} E \\ \vartheta \end{matrix} \right\rangle ; \left\langle \begin{matrix} E \\ \varepsilon \end{matrix} \middle| \dot{V}\begin{pmatrix} e \\ \varepsilon \end{pmatrix} \middle| \begin{matrix} E \\ \varepsilon \end{matrix} \right\rangle \end{pmatrix} = \begin{pmatrix} 0\ ;\ X_e/\sqrt{2} \\ X_e/\sqrt{2}\ ;\ 0 \end{pmatrix} = X_e C_2 \tag{A12}$$

$$\begin{pmatrix} \left\langle \begin{matrix} E \\ \vartheta \end{matrix} \middle| \dot{V}\begin{pmatrix} a \\ \bullet \end{pmatrix} \middle| \begin{matrix} E \\ \vartheta \end{matrix} \right\rangle ; \left\langle \begin{matrix} E \\ \vartheta \end{matrix} \middle| \dot{V}\begin{pmatrix} a \\ \bullet \end{pmatrix} \middle| \begin{matrix} E \\ \varepsilon \end{matrix} \right\rangle \\ \left\langle \begin{matrix} E \\ \varepsilon \end{matrix} \middle| \dot{V}\begin{pmatrix} a \\ \bullet \end{pmatrix} \middle| \begin{matrix} E \\ \vartheta \end{matrix} \right\rangle ; \left\langle \begin{matrix} E \\ \varepsilon \end{matrix} \middle| \dot{V}\begin{pmatrix} a \\ \bullet \end{pmatrix} \middle| \begin{matrix} E \\ \varepsilon \end{matrix} \right\rangle \end{pmatrix} = \begin{pmatrix} X_a\ ;\ 0 \\ 0\ ;\ X_a \end{pmatrix} = X_a C_1 \tag{A13}$$

with

$$C_1 = \begin{pmatrix} 1 & 0 \\ 0 & 1 \end{pmatrix} \quad \ldots a_1 \tag{A14}$$

$$C_2 = \frac{1}{\sqrt{2}} \begin{pmatrix} 0 & 1 \\ 1 & 0 \end{pmatrix} \quad \ldots e_\varepsilon \tag{A15}$$

$$C_3 = \frac{1}{\sqrt{2}} \begin{pmatrix} -1 & 0 \\ 0 & 1 \end{pmatrix} \ldots e_\vartheta \tag{A16}$$

This set may be completed by

$$C_4 = \frac{1}{\sqrt{2}} \begin{pmatrix} 0 & 1 \\ -1 & 0 \end{pmatrix} \quad \ldots a_2 \tag{A17}$$

Therefore,

$$V^{(1)} = C_2 X_e Q_{e\varepsilon} + C_3 X_e Q_{e\vartheta} \tag{A18}$$

is valid for groups in which the symmetrized direct product is $[E \otimes E] = A_1 + E$ (e.g. for cubic groups). For the remaining groups with $[E \otimes E] = A_1 + B_1 + B_2$ (e.g. for D_{4h}), the expression

$$V^{(1)} = C_2 X_{b_1} Q_{b_1} + C_3 X_{b_2} Q_{b_2} \tag{A19}$$

holds.

In the reduction of quadratic matrix elements V^{rs}_{kl}, the operator part $\ddot{V}\begin{pmatrix}\Gamma_r & \Gamma_s\\ \gamma_r & \gamma_s\end{pmatrix}$ represents a component of the basis set that is obtained by a direct product $\Gamma_r \otimes \Gamma_s$. Its representation is, in general, reducible. The decomposition of the operator part may be performed according to the Wigner formula

$$\left|\begin{matrix}\Gamma_r\\ \gamma_r\end{matrix}\right\rangle \left|\begin{matrix}\Gamma_s\\ \gamma_s\end{matrix}\right\rangle = \sum_{\Gamma_i,\gamma_i} \left\langle\begin{matrix}\Gamma_r & \Gamma_s\\ \gamma_r & \gamma_s\end{matrix}\middle|\begin{matrix}\Gamma_i\\ \gamma_i\end{matrix}\right\rangle |\Gamma_r\Gamma_s\Gamma_i\gamma_i\rangle \tag{A20}$$

where the summations run over all Γ_i representations and their γ_i components are contained in the direct product

$$\Gamma_r \otimes \Gamma_s = \Gamma_{red} = \sum_i \Gamma_i \tag{A21}$$

Therefore

$$\begin{aligned}\left\langle\begin{matrix}\Gamma_k\\ \gamma_k\end{matrix}\right| \ddot{V}\begin{pmatrix}\Gamma_r & \Gamma_s\\ \gamma_r & \gamma_s\end{pmatrix} \left|\begin{matrix}\Gamma_l\\ \gamma_l\end{matrix}\right\rangle &= \sum_{\Gamma_i,\gamma_i \in \Gamma_r\otimes\Gamma_s} \left\langle\begin{matrix}\Gamma_r & \Gamma_s & \Gamma_i\\ \gamma_r & \gamma_s & \gamma_i\end{matrix}\right\rangle \left\langle\begin{matrix}\Gamma_k\\ \gamma_k\end{matrix}\right| Y(\Gamma_r\Gamma_s\Gamma_i\gamma_i) \left|\begin{matrix}\Gamma_l\\ \gamma_l\end{matrix}\right\rangle = \\ &= X(\Gamma_k\Gamma_l\,;\,\Gamma_r\Gamma_s) \sum_{\Gamma_i,\gamma_i} \left\langle\begin{matrix}\Gamma_r & \Gamma_s\\ \gamma_r & \gamma_s\end{matrix}\middle|\begin{matrix}\Gamma_i\\ \gamma_i\end{matrix}\right\rangle \left\langle\begin{matrix}\Gamma_l & \Gamma_i\\ \gamma_l & \gamma_i\end{matrix}\middle|\begin{matrix}\Gamma_k\\ \gamma_k\end{matrix}\right\rangle\end{aligned} \tag{A22}$$

is valid.

For example, in cubic groups the reduction proceeds as

$$\begin{aligned}\left\langle\begin{matrix}E\\ \vartheta\end{matrix}\right| \ddot{V}\begin{pmatrix}e & e\\ \vartheta & \vartheta\end{pmatrix} \left|\begin{matrix}E\\ \vartheta\end{matrix}\right\rangle &= \sum_{\substack{\Gamma = a_1, a_2, e\\ \gamma_e = \vartheta,\varepsilon}} \left\langle\begin{matrix}e & e\\ \vartheta & \vartheta\end{matrix}\middle|\begin{matrix}\Gamma\\ \gamma\end{matrix}\right\rangle \left\langle\begin{matrix}E\\ \vartheta\end{matrix}\right| Y(ee\Gamma_\gamma) \left|\begin{matrix}E\\ \vartheta\end{matrix}\right\rangle = \\ &= \left\langle\begin{matrix}e & e\\ \vartheta & \vartheta\end{matrix}\middle|\begin{matrix}a_1\\ \bullet\end{matrix}\right\rangle \left\langle\begin{matrix}E\\ \vartheta\end{matrix}\right| Y(eea_1\bullet) \left|\begin{matrix}E\\ \vartheta\end{matrix}\right\rangle + \left\langle\begin{matrix}e & e\\ \vartheta & \vartheta\end{matrix}\middle|\begin{matrix}a_2\\ \bullet\end{matrix}\right\rangle \left\langle\begin{matrix}E\\ \vartheta\end{matrix}\right| Y(eea_2\bullet) \left|\begin{matrix}E\\ \vartheta\end{matrix}\right\rangle + \\ &+ \left\langle\begin{matrix}e & e\\ \vartheta & \vartheta\end{matrix}\middle|\begin{matrix}e\\ \vartheta\end{matrix}\right\rangle \left\langle\begin{matrix}E\\ \vartheta\end{matrix}\right| Y(eee\vartheta) \left|\begin{matrix}E\\ \vartheta\end{matrix}\right\rangle + \left\langle\begin{matrix}e & e\\ \vartheta & \vartheta\end{matrix}\middle|\begin{matrix}e\\ \varepsilon\end{matrix}\right\rangle \left\langle\begin{matrix}E\\ \vartheta\end{matrix}\right| Y(eee\varepsilon) \left|\begin{matrix}E\\ \vartheta\end{matrix}\right\rangle = \\ &= \left\langle\begin{matrix}e & e\\ \vartheta & \vartheta\end{matrix}\middle|\begin{matrix}e\\ \vartheta\end{matrix}\right\rangle \left\langle\begin{matrix}E & e\\ \vartheta & \vartheta\end{matrix}\middle|\begin{matrix}E\\ \vartheta\end{matrix}\right\rangle X(EE\,;\,ee) = \left(\frac{-1}{\sqrt{2}}\right)\left(\frac{-1}{\sqrt{2}}\right) X_{ee} = \tfrac{1}{2} X_{ee}\end{aligned} \tag{A23}$$

Capital letters denote the symmetry of electronic states and small letters that of vibrational modes.

The important finding is that the integral

$$\begin{aligned}
\left\langle \begin{matrix}E\\ \vartheta\end{matrix} \middle| V\begin{pmatrix}a_1 & e\\ \cdot & \vartheta\end{pmatrix} \middle| \begin{matrix}E\\ \vartheta\end{matrix}\right\rangle &= \left\langle \begin{matrix}a_1 & e\\ \cdot & \vartheta\end{matrix} \middle| \begin{matrix}e\\ \vartheta\end{matrix}\right\rangle \left\langle \begin{matrix}E\\ \vartheta\end{matrix} \middle| Y(a_1ee\vartheta) \middle| \begin{matrix}E\\ \vartheta\end{matrix}\right\rangle + \\
&+ \left\langle \begin{matrix}a_1 & e\\ \cdot & \vartheta\end{matrix} \middle| \begin{matrix}e\\ \varepsilon\end{matrix}\right\rangle \left\langle \begin{matrix}E\\ \vartheta\end{matrix} \middle| Y(a_1ee) \middle| \begin{matrix}E\\ \vartheta\end{matrix}\right\rangle = \\
&= \left\langle \begin{matrix}a_1 & e\\ \cdot & \vartheta\end{matrix} \middle| \begin{matrix}e\\ \vartheta\end{matrix}\right\rangle \left\langle \begin{matrix}E & e\\ \vartheta & \vartheta\end{matrix} \middle| \begin{matrix}E\\ \vartheta\end{matrix}\right\rangle X(EE\,;\,a_1e) = \\
&= 1 \cdot \left(-\frac{1}{\sqrt{2}}\right) X_{ae}
\end{aligned} \tag{A24}$$

and its analogues are non-zero, a fact which used to be overlooked.

Analogously, the cubic matrix element V_{kl}^{rst} is reduced as

$$\begin{aligned}
&\left\langle \begin{matrix}\Gamma_k\\ \gamma_k\end{matrix} \middle| \dddot{V}\begin{pmatrix}\Gamma_r & \Gamma_s & \Gamma_t\\ \gamma_r & \gamma_s & \gamma_t\end{pmatrix} \middle| \begin{matrix}\Gamma_l\\ \gamma_l\end{matrix}\right\rangle = \sum_{\Gamma_i\gamma_i \in \Gamma_r \bullet \Gamma_s} \left\langle \begin{matrix}\Gamma_r & \Gamma_s\\ \gamma_r & \gamma_s\end{matrix} \middle| \begin{matrix}\Gamma_i\\ \gamma_i\end{matrix}\right\rangle \left\langle \begin{matrix}\Gamma_k\\ \gamma_k\end{matrix} \middle| Z\left(\Gamma_r\Gamma_s\begin{bmatrix}\Gamma_i\\ \gamma_i\end{bmatrix} \begin{matrix}\Gamma_t\\ \gamma_t\end{matrix}\right) \middle| \begin{matrix}\Gamma_l\\ \gamma_l\end{matrix}\right\rangle = \\
&= \sum_{\Gamma_i\gamma_i \in \Gamma_r \bullet \Gamma_s} \left\langle \begin{matrix}\Gamma_r & \Gamma_s\\ \gamma_r & \gamma_s\end{matrix} \middle| \begin{matrix}\Gamma_i\\ \gamma_i\end{matrix}\right\rangle \sum_{\Gamma_j\gamma_j \in \Gamma_i \bullet \Gamma_t} \left\langle \begin{matrix}\Gamma_i & \Gamma_t\\ \gamma_i & \gamma_t\end{matrix} \middle| \begin{matrix}\Gamma_j\\ \gamma_j\end{matrix}\right\rangle \left\langle \begin{matrix}\Gamma_k\\ \gamma_k\end{matrix} \middle| Y\left(\Gamma_r\Gamma_s\Gamma_t\begin{bmatrix}\Gamma_j\\ \gamma_j\end{bmatrix}\right) \middle| \begin{matrix}\Gamma_l\\ \gamma_l\end{matrix}\right\rangle = \\
&= \sum_{\Gamma_i\gamma_i} \left\langle \begin{matrix}\Gamma_r & \Gamma_s\\ \gamma_r & \gamma_s\end{matrix} \middle| \begin{matrix}\Gamma_i\\ \gamma_i\end{matrix}\right\rangle \cdot \sum_{\Gamma_j\gamma_j} \left\langle \begin{matrix}\Gamma_i & \Gamma_t\\ \gamma_i & \gamma_t\end{matrix} \middle| \begin{matrix}\Gamma_j\\ \gamma_j\end{matrix}\right\rangle \cdot \left\langle \begin{matrix}\Gamma_l & \Gamma_j\\ \gamma_l & \gamma_j\end{matrix} \middle| \begin{matrix}\Gamma_k\\ \gamma_k\end{matrix}\right\rangle \cdot X(\Gamma_k\Gamma_l\,;\,\Gamma_r\Gamma_s\Gamma_t)
\end{aligned} \tag{A25}$$

13 Symbols and Abbreviations

a symmetry of vibration
A symmetry of electron state; vibronic constant
APS adiabatic potential surface
AOM Angular Overlap Model
b symmetry of vibration
B symmetry of electron state; vibronic constant
C_i matrix of coupling coefficients
CG Clebsh-Gordan coefficient
CNDO Complete Neglect of Differential Overlap
e symmetry of vibration
E symmetry of electron state; energy, eigenvalue of Schrödinger equation
f_k nuclear (vibration) function
F_i, F_{ij}, F_{ijk} ... pure nuclear term of force constants
G symmetry point group
$\hat{G}_{lk}$ nonadiabatic coupling operator
$\hat{H}$ Hamiltonian
H_{kl} matrix element of Hamiltonian
K_i, K_{ij}, K_{ijkl} ... force constants
LCAO Linear Combination of Atomic Orbitals
m_A mass of nucleus
MO Molecular Orbital
Σ group order
Q nuclear coordinate
r electron coordinate; radial (polar) coordinate; distance

$\hat{R}$	resolvent
$Sp(V)$	trace of matrix V
t	symmetry of vibration
T	symmetry of electron state
$\hat{T}$	Kinetic energy operator
U	interaction potential
$\hat{V}$	potential energy operator
$\dot{V}, \ddot{V}, \dddot{V} \ldots$	derivatives of $\hat{V}$ (first, second, third...)
X	reduced matrix element
$X_i, X_{ij}, X_{ijk} \ldots$	electron-nuclear term of force constants; vibronic constant
$Y_{j,m}$	spherical harmonics with m and j quantum numbers
Z	vibronic constant
γ	component of irreducible representation Γ
Γ	representation of symmetry
δ_{ij}	Kronecker delta
ε	eigenvalue of matrix; vibronic correction term; component of E representation
ϑ	angular coordinate; component of E representation
ϱ	radial coordinate, Jahn-Teller radius
φ	angular coordinate
ϕ	electronic wave function
∇	nabla operator

14 References

1. Murrell, J. N.: Struct. Bonding *32,* 93 (1977)
2. Ammeter, J. H.: Nouv. J. Chim. *4,* 631 (1980)
3. Ammeter, J. H., Bürgi, H. B. Gamp, E., Meyer-Sandrin, V., Jensen W. P.: Inorg. Chem. *18,* 733 (1979)
4. Gažo, J.: Pure Appl. Chem. *38,* 279 (1974)
5. Hathaway, B. J.: Coord. Chem. Rev. *35,* 211 (1981); ibid. *41,* 423 (1982); ibid. *52,* 87 (1983); Struct. Bonding *57,* 55 (1984)
6. Gažo, J., Bersuker, I. B., Garaj, J., Kabešová, M., Kohout, J., Langfelderová, H., Melnik, M., Serátor, M., Valach, F.: Coord. Chem. Rev. *19,* 253 (1976)
7. Hathaway, B. J., Duggan, M., Murphy, A., Mullane, J., Power D. C., Walsh, A., Walsh, B.: ibid. *36,* 267 (1981)
8. Gažo, J., Boča, R., Jóna, E., Kabešová, M., Macášková, Ľ., Šima, J., Pelikán, P., Valach, F.: ibid. *43,* 87 (1982)
9. Gažo, J., Boča, R., Jóna, E., Kabešová, M., Macášková, Ľ., Šima, J.: Pure Appl. Chem. *55,* 65 (1983)
10. Reinen, D.: Comments Inorg. Chem. *2,* 227 (1983)
11. Englman, R.: The Jahn-Teller Effect in Molecules and Crystals. Wiley-Interscience, New York 1972
12. Bersuker, I. B.: Coord. Chem. Rev. *14,* 357 (1975)
13. Bersuker, I. B., Polinger V. Z.: Adv. Quantum Chem. *15,* 85 (1982)
14. Bacci, M.: Nouv. J. Chim. *4,* 577 (1980)
15. Bacci, M.: Struct. Bonding *55,* 67 (1983)
16. Reinen, D., Friebel, C.: ibid. *37,* 1 (1979)
17. Pearson, R. G.: Symmetry Rules for Chemical Reactions. Wiley-Interscience, New York 1976
18. Özkan, I., Goodman, L.: Chem. Rev. *79,* 275 (1979)
19. Lowe, J. P.: Quantum Chemistry. Academic Press, New York 1978
20. Jahn, H. A., Teller, E.: Proc. Roy. Soc. *A 161,* 220 (1937); Jahn, H. A.: Proc. Roy. Soc. *A 164,* 117 (1938)
21. Salthouse, J. A., Ware, M. J.: Point Group Character Tables and Related Data. University Press, Cambridge 1972
22. Ruch, E., Schönhofer, A.: Theoret. Chim. Acta *3,* 291 (1965)
23. Blount E. I.: J. Math. Phys. *12,* 1890 (1971)
24. Pelikán, P., Breza, M., Boča, R.: Polyhedron *4,* 1543 (1985)
25. Breza, M., Pelikán, P., Boča, R.: ibid. *5,* 1607 (1986)
26. Pelikán, P., Breza, M., Boča, R.: ibid. *5,* 753 (1986)

27. Coffman, R. E.: J. Chem. Phys. *44,* 2305 (1966)
28. Öpik, U., Pryce, M. H. L.: Proc. Roy. Soc. *A 238,* 425 (1957)
29. O'Brien, M.: ibid. *A 281,* 323 (1964)
30. Goodenough, J.: J. Phys. Chem. Solids *25,* 151 (1964)
31. Boča, R., Pelikán, P., Breza, M., Gažo, J.,: Polyhedron *2,* 921 (1983)
32. Judd, B. R.: Can. J. Phys. *52,* 999 (1974)
33. Jotham, R. W., Kettle, S. F. A.: Inorg. Chim. Acta *5,* 183 (1971)
34. Dagis, R. S., Levinson, I. B.: in: Frish, S. E. (ed.) Optika i spektroskopiya 3. Molekulyarnaya spektroskopiya. Nauka, Moscow 1967
35. Pelikán, P., Breza, M.: Chem. Papers *39,* 255 (1985)
36. Pelikán, P., Breza, M.: J. Mol. Struct. (Theochem.) *124,* 231 (1985)
37. Liehr, A. D.: J. Phys. Chem. *67,* 389 (1963)
38. Boča, R.: Chem. Papers *35,* 769 (1981)
39. Boča, R.: ibid. *35,* 779 (1981)
40. Boča, R.: ibid. *37,* 297 (1983)
41. Cullen, D. L., Lingafelter, E. C.: Inorg. Chem. *10,* 1265 (1971); Noda, Y., Mori, M., Yamada, Y.: Solid State Commun. *19,* 1071 (1976); Joesten, M. D., Takagi, S., Lenhert, P. G.: Inorg. Chem. *16,* 2681 (1977)
42. Hellwege, K. H. (ed.): Crystal Structures of Inorganic Compounds. Key-Elements: d^9-, d^{10}-, $d^1 \ldots d^3$-, f-Elements (Landolt-Börnstein. New Series. Vol. III/7 e). Springer, Berlin – Heidelberg – New York 1976
Hellwege, K. H. (ed.): Crystal Structures of Inorganic Compounds. Key-Elements: $d^4 \ldots d^8$-elements (Landolt-Börnstein. New Series. Vol. III/7 f). Springer, Berlin – Heidelberg – New York 1977
43. van Vleck, J. H.: J. Chem. Phys. *7,* 1972 (1939)
44. Bersuker, I. B.: Elektronnoe stroenie i svoistva koordinatsionnykh soedinenii. Khimiya, Leningrad 1976
45. Bersuker, I. B.: Zh. Eksp. Teor. Fiz. *43,* 1315 (1962)
46. Bacci, M.: Chem. Phys. *40,* 237 (1979)
47. Bacci, M.: Biophys. Chemistry *11,* 39 (1980)
48. Warren, K. D.: Struct. Bonding *57,* 119 (1984)
49. Nikiforov, A. E., Shashkin, S. Yu., Krotkii, A. I.: Phys. Stat. Sol. (b) *97,* 475 (1980)
50. Nikiforov, A. E., Shashkin, S. Yu., Krotkii, A. I.: ibid. *98,* 289 (1980)
51. Lee, T. J., Fox, D. J., Shaeffer III, H. F., Pitzer, R. M.: J. Chem. Phys. *81,* 356 (1984)
52. Dixon, R. N.: Mol. Phys. *20,* 113 (1971)
53. Pelikán, P., Breza, M., Liška, M.: Inorg. Chim. Acta *45,* L 1 (1980)
54. Griffith, J. S.: The Theory of Transition-Metal Ions. University Press, Cambridge 1964
55. Griffith, J. S.: The Irreducible Tensor Method for Molecular Symmetry Groups. Prentice-Hall, Englewood Cliffs NJ 1962
56. Jucys, A. P., Bandzaitis, A. A.: Theory of Angular Momentum in Quantum Mechanics (in russian). Mokslas, Vilnius 1977
57. Hellwege, K. H. (ed.): Numerical Tables for Angular Correlation Computations . . . (Landolt-Börnstein. New Series. Vol. I/3). Springer, Berlin – Heidelberg – New York 1968

A Dynamic Ligand Field Theory for Vibronic Structures Rationalizing Electronic Spectra of Transition Metal Complex Compounds

Hans-Herbert Schmidtke and Joachim Degen

Institut für Theoretische Chemie der Universität Düsseldorf, Universitätsstraße 1, D-4000 Düsseldorf 1, F.R.G.

A ligand field model which considers the dynamics of vibrating ligands of transition metal ions is formulated, maintaining fundamental assumptions and model parameters of common (static) ligand field theory. It allows a survey to be obtained on corresponding theoretical approaches available on various occasions in the literature which interpret vibrational fine structures of band progressions in electronic absorption and emission spectra including the entanglements caused by Jahn-Teller effects. In the unified model the electronic d-states are vibronically coupled to the nuclei moving in a harmonic force field applying first order perturbation terms in the Herzberg-Teller approximation. The linear vibronic coupling constants, stabilization energies and geometric distortions due to excitation can be calculated from appropriate ligand field parameters of the ground state which are obtained by comparison with the measured spectra and from metal-ligand atomic equilibrium distances. The model is applied to systems of octahedral symmetry representing d-electron coupling with α_1, ε and/or τ_2 vibrations for which well resolved vibronic spectra have been reported.

1 Introduction 100

2 Theory 101
2.1 Potential Energy Derivatives 102
2.2 Vibronic Coupling Arising from α_1 Vibrations 104
2.2.1 Coupling to Single-Electron (Orbital) Levels 104
2.2.2 Coupling to Many-Electron States 109
2.3 Vibronic Coupling Arising from ε and τ_2 Vibrations 112
2.3.1 Orbital Vibronic Couplings 112
2.3.2 ε and τ_2 Coupling to Many-Electron States 114

3 Application to Vibronic Spectra 117
3.1 Vibronic Analysis by Parameter Adaptation 117
3.2 Parameter Calculations by the Dynamic Ligand Field Model 119
3.2.1 α-Coupling Cases 119
3.2.2 ε-Coupling Cases 121

4 Conclusion 122

5 References 123

© Springer-Verlag Berlin Heidelberg 1989

1 Introduction

Common ligand field theory calculates the electronic structure of complex compounds of open shell transition or rare earth metal ions of low lying electronic states. The nuclear framework is assumed to be fixed and transitions between energy levels occur at nuclear equilibrium positions which do not change during the transition. A model which is limited by considering the electronic structure can only describe properly transitions between potential minima which are at equal positions for both the ground and excited states. Transitions between potential surface minima shifted along one of the nuclear coordinates, if the Franck-Condon principle is obeyed, must always involve a change in the vibrational excitation which, however, is not taken into account satisfactorily by a purely electronic model. As a result, the ligand field parameters derived from the experiment contain in addition to the energy gap between two electronic levels at minimum potentials, also various vibrational quanta which are excited during the transition (cf. Fig. 1 for the first spin-allowed transition in an octahedral d^3-complex). A proper theory for rationalizing vibronic spectra therefore should include also vibrations around the average positions of the nuclei by considering the interaction between electrons and the moving nuclei which is generally referred to as vibronic coupling. The electrostatic model then must be extended towards a dynamic theory which involves also the motion of the nuclei.

The theory introduces the molecular dynamics in general by applying a Herzberg-Teller coupling scheme in which the electronic Hamiltonian depending on the symmetry adapted nuclear coordinates S_i of the molecule is expanded around the equilibrium position S^0 into a Taylor series

$$H(q, S) = H(q, S^0) + \sum_{i=1}^{3N-6} \left(\frac{\partial H}{\partial S_i}\right)_{S^0} S_i + \frac{1}{2} \sum_{i,j=1}^{3N-6} \left(\frac{\partial^2 H}{\partial S_i \partial S_j}\right)_{S^0} S_i S_j + \ldots \tag{1}$$

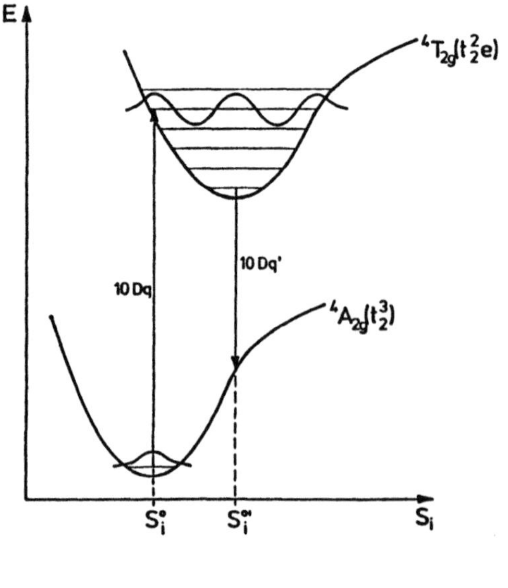

Fig. 1. Franck-Condon transitions between shifted potential energy curves in the case of octahedral d^3 systems supplying different ligand field parameters for the ground and the excited state

where q and S denote sets of electronic and nuclear coordinates, respectively, and N the number of atoms in the molecule[1-3]. The zero order term $H(q, S^0)$ represents the electronic Hamiltonian at fixed nuclear coordinates S^0 in the adiabatic approximation which allows the total wavefunction to be written as a product

$$\Psi(q, S) = \Psi(q, S^0) \cdot \chi(S) \,, \tag{2}$$

leading to the crude Born-Oppenheimer separation of electron and nuclear coordinates. The original electrostatic ligand field theory is founded on the approximation represented by $H(q, S^0)$ of Eq. (1) with a potential part containing the ligand field operator $V_{LF}(q, S^0)$ which considers the interaction of valence electrons of the central ion with a static coulomb field arising from the ligands at fixed position S^0. In the semiempirical theory the electrostatic viewpoint is abandoned by collecting all metal-ligand interaction into parameters which are determined from the experiment leaving only the structural part (symmetry, valence angles) in the energy formulas to the calculation. A dynamic ligand field theory then must include the additional terms in Eq. (1) which account for the vibronic coupling of the ligand field levels to the coordinates S_i of the nuclei where N refers to the number of ligands in the complex molecule[1]. It will be shown that in this case the electrostatic model also can be transferred into a semiempirical one by introducing the same type of parameters known for the static ligand field model which are determined from the experiment.

In the following we consider the first order dynamic ligand fields of all possible nuclear vibrations which give finite contributions to d-orbital perturbations taking octahedrally coordinated transition metal ions as example. For this, the derivatives of the ligand field operator with respect to nuclear symmetry coordinates are needed. The coupling matrices for degenerate vibrations are those pertinent to the Jahn-Teller effect[2-4] and the corresponding coupling constants will be calculated in terms of ligand field parameters and bond distances determined from the experiment. The model is applied to systems in which some of the possible coupling cases are realized. From the results stabilization energies and geometry distortions in the excited states due to vibronic interactions can be calculated.

2 Theory

Since vibronic coupling effects in the adiabatic approximation are generally small compared to energy differences between electronic levels a first order perturbation can be applied. The perturbation operator contains all terms depending on the nuclear coordinates S_i in the Herzberg-Teller series of Eq. (1) acting on the system with fixed nuclei which is represented by the zero order term which in ligand field theory is

$$H(q, S^0) = H_c(q, S_c^0) + V_{LF}(q, S^0) \tag{3}$$

where the Hamiltonian $H_c(q, S_c^0)$ of the central metal ion and the ligand field potential $V_{LF}(q, S^0)$ are taken at equilibrium coordinates of the central ion nucleus S_c^0 and the ligand nuclei S^0. The unperturbed functions $\Psi(q, S^0)$ are the solutions of a conventional ligand field calculation obtained from the Hamiltonian $H(q, S^0)$ yielding energy levels for the complex molecules of various symmetry types determined by the ligand field operator $V_{LF}(q, S^0)$. For central ions with only one d-electron the resulting functions are ligand field orbitals belonging to irreducible representations of the corresponding point group. In the case of molecules with octahedral symmetry d-orbitals are split into e and t_2 sets interacting with ligand orbitals by σ- and π-overlap, respectively. The energy difference between e and t_2 orbitals is by convention the ligand field parameter

$$10\,Dq = E(e) - E(t_2) \tag{4}$$

which in the semiempirical theory can be determined by experiment (for instance from the absorption spectra) and is valid at equilibrium position S^0 for the electronic ground state[5,6]. The potential curve of the excited states, on the other hand, may not have their minimum at the same nuclear position as for the ground state. For these internuclear distances the parameter Dq should be different (cf. Fig. 1), it can be determined, e.g., from excited state spectroscopy[7].

2.1 Potential Energy Derivatives

In the perturbation treatment we shall consider in the first place operators in Eq. (1) which are linear in S_i containing derivatives $\frac{\partial V_{LF}(q, S)}{\partial S_i}$. Since correlation to metal-ligand vibrations is intended the nuclear coordinates S_i are chosen as normal coordinates which are either equal to symmetry coordinates or can be written as linear combinations of the latter if an irreducible representation belonging to nuclear motions occurs several times. For systems with a center of inversion as octahedral molecules only even (gerade) coordinates may cause shifts of ligand field potential energy curves which is due to symmetry reasons imposed on the expectation values of these operator derivatives. Therefore only coordinates of even symmetry need to be considered from the set of vibrational coordinates S_i. Since for even modes the central atom is not involved in the vibrational motion it does not contribute to the perturbation. The derivatives in the linear vibronic coupling operators of Eq. (1) then reduce to $\frac{\partial H}{\partial S_i} = \frac{\partial V_{LF}}{\partial S_i}$ for even S_i coordinates. Since V_{LF} is usually given in terms of spherical harmonics depending on polar coordinates R_k, Θ_k, Φ_k of the ligands[5,6], their derivatives are calculated from derivatives with respect to vibrational elongations X_k, Y_k, Z_k of the ligand k, e.g. in the case of X_k

$$\frac{\partial V_{LF}}{\partial X_k} = \frac{\partial V_{LF}}{\partial R_k}\frac{\partial R_k}{\partial X_k} + \frac{\partial V_{LF}}{\partial \Theta_k}\frac{\partial \Theta_k}{\partial X_k} + \frac{\partial V_{LF}}{\partial \Phi_k}\frac{\partial \Phi_k}{\partial X_k} \tag{5}$$

These allow the desired potential derivatives with respect to symmetry coordinates to be calculated (N is the number of the ligands) from

$$\frac{\partial V_{LF}}{\partial S_i} = \sum_{k=1}^{N}\left(\frac{\partial V_{LF}}{\partial X_k}\frac{\partial X_k}{\partial S_i} + \frac{\partial V_{LF}}{\partial Y_k}\frac{\partial Y_k}{\partial S_i} + \frac{\partial V_{LF}}{\partial Z_k}\frac{\partial Z_k}{\partial S_i}\right) \tag{6}$$

Table 1. Ligand field potential energy derivatives $\left\langle -\left(\frac{\partial V_{LF}}{\partial S_i}\right)_{S^0}\right\rangle_r$ with respect to even octahedral symmetry coordinates S_i integrated over the radial parts in terms of different powers of λ[a]

i	S_i[b]	Symmetry	$\lambda = 0$	$\lambda = 2$	$\lambda = 4$
1	$\sqrt{\frac{1}{6}}(X_1 + Y_2 + Z_3 - X_4 - Y_5 - Z_6)$	α_1	$\varepsilon_0/\sqrt{6}R$	0	$70\sqrt{\frac{\pi}{6}}\frac{Dq}{R}\left[Y_4^0 + \sqrt{\frac{5}{14}}(Y_4^4 + Y_4^{-4})\right]$
2	$\sqrt{\frac{1}{12}}(-X_1 - Y_2 + 2Z_3 + X_4 + Y_5 - 2Z_6)$	ε	0	$\frac{108}{5}\sqrt{\frac{5\pi}{3}}\frac{Dq}{R\eta}Y_2^0$	$25\sqrt{\frac{\pi}{15}}\frac{Dq}{R}\left[\sqrt{5}\,Y_4^0 - \sqrt{\frac{7}{2}}(Y_4^4 + Y_4^{-4})\right]$
3	$\frac{1}{2}(X_1 - Y_2 - X_4 + Y_5)$	ε	0	$\frac{108}{5}\sqrt{\frac{5\pi}{6}}\frac{Dq}{R\eta}[Y_2^2 + Y_2^{-2}]$	$-50\sqrt{\frac{\pi}{10}}\frac{Dq}{R}[Y_4^2 + Y_4^{-2}]$
4	$\frac{1}{2}(Y_1 + X_2 - Y_4 - X_5)$	τ_2	0	$24\sqrt{\frac{3\pi}{10}}\frac{Dq}{R\eta}i[Y_2^2 - Y_2^{-2}]$	$-4\sqrt{\frac{5\pi}{2}}\frac{Dq}{R}i[Y_4^2 - Y_4^{-2}]$
5	$\frac{1}{2}(Z_1 + X_3 - Z_4 - X_6)$	τ_2	0	$24\sqrt{\frac{3\pi}{10}}\frac{Dq}{R\eta}[Y_2^1 - Y_2^{-1}]$	$\sqrt{5\pi}\frac{Dq}{R}[(Y_4^1 - Y_4^{-1}) + \sqrt{7}(Y_4^3 - Y_4^{-3})]$
6	$\frac{1}{2}(Z_2 + Y_3 - Z_5 - Y_6)$	τ_2	0	$-24\sqrt{\frac{3\pi}{10}}\frac{Dq}{R\eta}i[Y_2^1 + Y_2^{-1}]$	$-\sqrt{5\pi}\frac{Dq}{R}i[(Y_4^1 + Y_4^{-1}) - \sqrt{7}(Y_4^3 + Y_4^{-3})]$

[a] Spherical harmonics Y_λ^κ as defined in Ref. 6

[b] Numbering of ligands positions are 1, 2, 3 on the positive x, y, z axis, respectively, and 4, 5, 6 on the negative axis

For the use in Eq. (1), the expressions have to be taken at equilibrium positions R_k^0, Θ_k^0, Φ_k^0 of the ligands. The results obtained for even vibrational coordinates of an octahedron (N = 6) are compiled in Table 1 when usual ligand field parameters

$$Dq = \frac{1}{6} Ze^2\overline{r^4}/R^5$$

$$\varepsilon_0 = 6\, Ze^2/R$$

$$\eta = \overline{r^4}/(\overline{r^2}R^2) \qquad (7)$$

are introduced replacing the point charge model. In Eq. (7) Z is the formal charge on the ligands, R the central metal-ligand distance and $\overline{r^n}$ refers to the mean value of the n-th power electron position operator calculated over the radial parts of the d-functions[5,6]. The expression for α-coupling is easily obtained by calculation of $\partial V_{LF}(R, \Theta, \Phi)/\partial R$, the others may be taken from Van Vlecks Table[8] if the different choice of ε coordinates S_2 and S_3 is observed which we introduce as currently used[2]. Since all modes of Table 1 belong to different irreducible representations the symmetry coordinates are identical with normal coordinates. Higher terms in the expansions of V_{LF} with powers $\lambda > 4$ need not to be considered because all these matrix elements calculated from d-orbital sets vanish by symmetry reasons.

Matrix elements of corresponding ligand field potential derivatives with respect to odd vibrational coordinates are important, on the other hand, for calculating intensities of vibrational induced d-d transitions[1]. Since electronic transitions between d-states are strictly parity forbidden in centrosymmetric systems the occurrence of these transitions must be explained by intermixing of odd charge transfer states into ligand field states by vibronic coupling via odd vibrational modes (promoting or active modes). Odd potential derivatives necessary for calculating off-diagonal matrix elements between these states are, however, not of interest in the present context.

2.2 *Vibronic Coupling Arising from α_1 Vibrations*

Since the vibronic coupling contributions to the hamiltonian are one-electron operators, as are the ligand field operators V_{LF}, the calculation of perturbation energies is relatively simple. We start by considering d-orbital levels which are subject to a first order coupling perturbation due to the interaction with a totally symmetric vibrational mode α_1 with the atoms moving along the coordinate S_1.

2.2.1 *Coupling to Single-Electron (Orbital) Levels*

For simplicity we treat in detail a one electron system d^1 which in octahedral symmetry has an e and t_2 energy level (unperturbed system). Although this case is not very realistic, since the Jahn-Teller effect being operative in both states will distort the system to lower symmetry, we want to follow the procedure as used in the strong field version of ligand field theory. The results will supply all necessary informations for calculating also larger systems. The first order perturbation energies are obtained from the α-coupling operator of Table 1 by calculating matrix elements

$$\Delta E(e) = \left\langle \psi_e \left| \left(\frac{\partial V_{LF}}{\partial S_1} \right)_{S^0} S_1 \right| \psi_e \right\rangle = -\frac{1}{\sqrt{6}\,R} [\varepsilon_0(R) + 30\,Dq(R)] S_1 = a^e S_1$$

$$\Delta E(t_2) = \left\langle \psi_{t_2} \left| \left(\frac{\partial V_{LF}}{\partial S_1} \right)_{S^0} S_1 \right| \psi_{t_2} \right\rangle = -\frac{1}{\sqrt{6}\,R} [\varepsilon_0(R) + 20\,Dq(R)] S_1 = a^t S_1 \quad (8)$$

where the ψ_i are electronic functions at equilibrium in the Born-Oppenheimer approximation, Eq. (2), which in the present case are molecular orbitals derived from d_{z^2} or $d_{x^2-y^2}$ for e and from d_{xy}, d_{xz} or d_{yz} for t_2 levels. The Dirac bracket symbol refers to integration over electron coordinates only and the a^i are substitutions introduced for the orbital coupling coefficients belonging to S_1 which have been denoted by "orbital vibronic constants" [2)]. Corresponding off-diagonal matrix elements of the totally symmetric potential derivative operator all vanish by symmetry reason. The parameters ε_0 and Dq given by Eq. (7) are functions of R. From optical spectra using static ligand field theory only the latter parameter can be determined from the experiment yielding parameter values Dq and Dq′ which refer to the equilibrium distance R of the electronic ground and R′ of the excited state, respectively[7)].

For the total energy including vibronic coupling depending on the nuclear coordinate, also restoring forces must be considered which, in general, are assumed to be harmonic yielding

$$E_e = E^0 + \frac{1}{2} k_\alpha^e S_1^2 + a^e S_1$$

$$E_t = \frac{1}{2} k_\alpha^t S_1^2 + a^t S_1 \quad (9)$$

Here E^0 denotes the electronic energy difference of the t_2 and e orbitals and the k_α^p are the force constants of these levels belonging to the α_1 vibration. The $E(S_1)$ functions in Eq. (9) are then parabolic curves with their minima shifted due to the linear term towards larger values of the S_1 coordinate since both coupling constants a_α^p are in general negative. This is seen, e.g., from a point charge model by inserting the functional expressions for $\varepsilon_0(R)$ and Dq(R) into Eq. (8) and using $\overline{r^4} < R^4$ because of $\bar{r} < R$ (also considering molecular orbitals will lead to the same result due to the larger σ-antibonding in e- than π-antibonding in t_2-orbitals).

The relative positions of potential curves are illustrated in Fig. 1. The restoring forces have their origin from the purely quadratic terms in the expansion series of Eq. (1), however, since the Hamiltonian in the derivatives, by definition, describes only the electronic energy they do not entirely correspond to the quadratic terms of Eq. (9) because the latter include also repulsions of the nuclear cores. On the other hand, when considering differences of restoring forces between excited and ground states by which contributions due to core repulsions are cancelled, the result can be related to corresponding term differences of Eq. (1) if, e.g., expectation values as

$$\left\langle \psi_e \left| \left(\frac{\partial^2 H}{\partial S_i^2} \right)_{S^0} \right| \psi_e \right\rangle S_i^2 - \left\langle \psi_{t_2} \left| \left(\frac{\partial^2 H}{\partial S_i^2} \right)_{S^0} \right| \psi_{t_2} \right\rangle S_i^2 \quad (10)$$

calculated at equilibrium position S^0 are considered.

Since normal coordinates are measured from equilibrium positions of the nuclei taking, in general, the electronic ground state as origin for these coordinates, we perform a transformation of the parabolic equations such that the linear term in Eq. (9) which refers to the lower energy level vanishes. Since for octahedral coordination E_t is always lower than E_e we therefore introduce

$$Q_1 = S_1 + \frac{a^t}{k_\alpha^t} \tag{11}$$

leading to the potential energy functions

$$\begin{aligned} E'_e &= E^{0\prime} - \Delta E^e + \tfrac{1}{2} k_\alpha^e Q_1^2 + a_\alpha Q_1 \\ E'_t &= \tfrac{1}{2} k_\alpha^t Q_1^2 \end{aligned} \tag{12}$$

in which the energy stabilization due to vibronic coupling in the lower state has been incorporated into the constant $E^{0\prime}$ (cf. Fig. 2)

$$E^{0\prime} = E^0 + \Delta E^e + \frac{a^t}{k_\alpha^t}\left[\frac{1}{2} a^t\left(1 + \frac{k_\alpha^e}{k_\alpha^t}\right) - a^e\right] \tag{13}$$

and a_α is a short hand notation for

$$a_\alpha = a^e - \frac{k_\alpha^e}{k_\alpha^t} a^t \tag{14}$$

$E^{0\prime}$ is identical to the zero phonon transition energy E_{0-0} related to the α-vibration except for a correction by corresponding zero point energies:

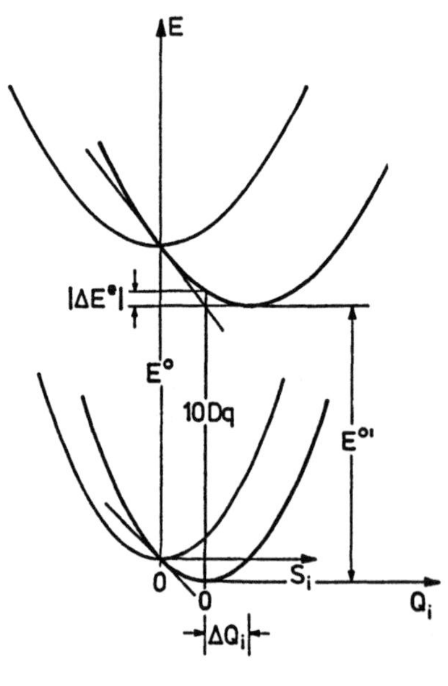

Fig. 2. Energy shifts of potential curves due to vibronic coupling and choice of vibrational coordinates

$$E^{0\prime} = E_{0-0} + \tfrac{1}{2}\hbar(\omega_\alpha^t - \omega_\alpha^e) \tag{15}$$

If E_{0-0} is identified with the zero phonon transition from the experiment that includes zero point energy effects of the other (non active) modes[1)], the extra terms are included into $E^{0\prime}$ which also is determined from the spectra fitting the vertical transition energy to $E^{0\prime} - \Delta E^e$ (vide infra). The shift of the excited state potential minimum with respect to that of the ground state, i.e. the equilibrium geometry distortion, is calculated from the parabolic formulas to be

$$\Delta Q_1 = -a_\alpha/k_\alpha^e \tag{16}$$

The stabilization energy due to the linear vibronic coupling in the e state with respect to the t_2 ground state is obtained from the E_e' potential function of Eq. (12) by insertion of ΔQ_1 for Q_1

$$\Delta E^e = -\frac{1}{2} k_\alpha^e \Delta Q_1^2 = -\frac{1}{2}\frac{a_\alpha^2}{k_\alpha^e} \tag{17}$$

The Eq. (13) and (17), since their parameters refer to the ground state equilibrium configuration of the nuclei $Q_i^0 = 0$ for any coordinate i, then can be put into relation to the ligand field parameter 10 Dq derived from the ground state spectra by

$$10\,Dq = E^{0\prime} + |\Delta E^e| \tag{18}$$

The equations may be verified by inspection of Fig. 2 where the notations made so far are explained by illustration. The force constants can also be determined from the experiment when substituting $k = m\omega^2$ where m is the effective mass and ω the angular frequency of the corresponding normal mode which for the ground state are obtained by carrying out, in general, a normal coordinate analysis on the vibrational (Raman and/or infrared) spectra. For even stretching modes (α_1 and ε) the reduced masses for both the ground and excited states are simply the masses of the ligands, i.e. $m_\alpha = m_\varepsilon = m(L)$. In the case of τ_2 coupling, which, however, has not been observed so far exhibiting vibrational progression, a normal coordinate analysis would have to be carried out. For excited states, values for ω are calculated from the frequency factor $\beta = \omega^e/\omega^g$ (relating respective excited state to ground state frequencies) which also can be determined from vibronic spectra[9)]. From the analysis of intensity distributions in vibrational progressions another parameter Δ_v is determined which is connected to the distortion parameter ΔQ_i, cf. Eq. (16), by

$$\Delta_v = (m_v\omega_v/\hbar)^{1/2}\Delta Q_i \tag{19}$$

which is valid for all Q_i belonging to α_1, ε and τ_2 vibrations[9)]. With the parameters ω, β and Δ the geometry distortion ΔQ_1, the stabilization energy ΔE due to vibronic coupling and the coupling constant a_α of Eq. (12) can be determined from the experiment using Eq. (16), (17) and (19). Furthermore by Eq. (15) and (18) the zero phonon energy E_{0-0} is obtained from the ligand field parameter Dq. The individual coupling constants a^e and a^t of Eq. (9), however, cannot be determined separately by this procedure since these are defined by Eq. (8) with respect to $S_1 = S_1^0 = 0$ which is not at equilibrium position of the ground or the excited state (cf. Fig. 2) such that Dq(R) cannot be obtained from the experiment. Also a transformation by Eq. (11) into the equilibrium position at $Q_1 = 0$ would again have to make use of the knowledge of a^t.

However, if the expansion of Eq. (1) is carried out with respect to coordinates, e.g., at ground state equilibrium position of the nuclei pertinent to Q_i coordinates, the metal-ligand distance R and the parameter functions $\varepsilon_0(R)$ and Dq(R) in the equations refer to the ground state and can be obtained by experiments carried out for this state. Since for minimal energy the linear term in Eq. (9) in the ground state must disappear, i.e. $a^t = 0$, we obtain from Eq. (8)

$$\varepsilon_0 = 20\,Dq \tag{20}$$

and insertion in

$$\Delta E(e) = -\frac{50}{\sqrt{6}R}\,Dq\,Q_1 \tag{21}$$

yields by comparison with Eq. (12)

$$a_\alpha = -\frac{50}{\sqrt{6}R}\,Dq \tag{22}$$

the latter relation being already given by Wilson and Solomon[10]. With these equations the coupling coefficient a_α and the zeroth order parameter ε_0 are subject to calculation from the ligand field parameter Dq and molecular equilibrium distance R both available from the experiment. We know that ε_0 cannot be determined from common ligand field formulas by comparison with electronic transitions; it obtains a natural explanation by the present procedure.

If the ligand field parameter Dq′ of the relaxed (at equilibrium) excited state is available from the evaluation of excited state spectra[7] a similar set of formulas can be derived. The ligand field parameters for the ground state and excited state equilibrium geometry Dq and Dq′, respectively, (cf. Fig. 1) are related for the one-electron case d^1 using the parabolic formula and Eq. (17) by

$$10\,Dq' = 10\,Dq + 2\,\Delta E^e = 10\,Dq - \frac{a_\alpha^2}{k_\alpha} \tag{23}$$

in which, for simplicity, equal force constants for the e and t_2 levels are assumed. The calculation of coupling constants and stabilization energies related to the excited state is then straightforward. We prefer to use, however, ground state ligand field parameters since these are much easier available from the experiment and will use normal coordinates Q_i of the ground state.

Using a quite similar model, Hitchman[11] has worked out a formula for the geometry distortion ΔQ_1 due to α_1-coupling if Dq varies with R^{-n} which in our notation is

$$\Delta Q_1 = \frac{n \cdot 10\,Dq}{\sqrt{N}k_\alpha R} \tag{24}$$

where N is the number of coordinated ligands (because valid only for cubic symmetry N is, e.g., 4 in T_d or 6 for O_h). Since he assumes n = 5 as in Eq. (7) this formula is identical with Eq. (16) when the coupling constant of Eq. (22) is introduced. In this derivation one uses the dependence of Dq′ on R′ = R + ΔR which is

$$Dq' \sim \frac{1}{(R + \Delta R)^n} = \frac{1}{R^n}\left(\frac{R + \Delta R}{R}\right)^{-n} \tag{25}$$

such that

$$Dq' = Dq\left(1 + \frac{\Delta R}{R}\right)^{-n} = Dq\left(1 - \frac{n\Delta R}{R}\right) \tag{26}$$

the latter being an approximation for $\Delta R \ll R$. With Eq. (17) and (23) we obtain considering the normalization $\Delta R = \Delta Q_1/\sqrt{6}$

$$\Delta Q_1 = \frac{n \cdot 10\ Dq}{\sqrt{6}\, k_\alpha R} \tag{27}$$

which is Eq. (24) for octahedral coordination N = 6.

2.2.2 *Coupling to Many-Electron States*

In the many electron case, where the electronic system of the central ion has a d^n configuration, the ligand field operator and its derivatives are sums over the number of d-electrons and the wave functions Ψ describe many electron states (multiplets) which transform according to irreducible representations Γ_p of the symmetry point group O_h. Then the perturbation of a Γ_p state due to coupling to the vibration α_1 is given by

$$\Delta E(\Gamma_p) = \left\langle \Psi_{\Gamma_p} \middle| \sum_1^n \left(\frac{\partial V_{LF}^{(1)}}{\partial S_1}\right)_{S^0} S_1 \middle| \Psi_{\Gamma_p} \right\rangle = A_\alpha^p S_1 \tag{28}$$

in which the many electron coupling constant A_α^p when resolving the matrix element into one electron integrals can be expressed by orbital vibronic constants a_α^p. Since these matrix elements are finite by symmetry reason for all possible levels Γ_p, a vibronic coupling with α_1 by linear terms in S_1 is expected for every Γ_p. Off diagonal elements between functions belonging to states of identical irreducible representation vanish in the strong field representation since also the octahedral LF operator $\sum_l V_{LF}^{(l)}$, which depends on the electronic coordinates in the same way as its derivatives with respect to S_1 (cf. Table 1), has no off-diagonal elements in the perturbation matrix.

Since in a system with a d^n electron configuration, in principle, transitions between all states $\Gamma_p \leftrightarrow \Gamma_q$ may occur, every one of these combinations must be considered. Due to vibronic coupling both the potential curves for the ground and excited state are shifted along nuclear coordinates, the corresponding coupling parameter A_α^p being generally different for each of these states which are represented by the wave function Ψ_{Γ_p} in Eq. (28). Since the coupling operator is a sum of operators each depending on one electron only, the perturbation expression of Eq. (28) for a state Γ_p resulting from a t_2^n configuration is calculated by the matrix element of Eq. (8) yielding

$$\Delta E(\Gamma_p \text{ from } t_2^n) = -\frac{n}{\sqrt{6}R}[\varepsilon_0(R) - 20\,Dq(R)]S_1 = na_\alpha^t S_1 = A_\alpha^p S_1 \tag{29}$$

Therefore, potential energy curves are stabilized for any of the states Γ_p belonging to the t_2^n configuration by an equal amount if interaction with levels resulting from configurations including occupation of e-orbitals is neglected. For Γ_p states which belong to $t_2^{n-m}e^m$ configurations the coupling coefficients of Eq. (8) apply as well. The corresponding perturbation energy neglecting again the interaction with other electron configurations is

$$\Delta E(\Gamma_p \text{ from } t_2^{n-m}e^m) = [(n-m)a_\alpha^t + ma_\alpha^e]S_1 = A_\alpha^p S_1 \tag{30}$$

By setting, as earlier, the coupling coefficient of Eq. (29) equal to zero if t_2^n is the ground state configuration, Eq. (20) which relates ε_0 and Dq, is obtained also in this case. If the ground state belongs to a $t_2^{n-m}e^m$ configuration the corresponding relation is calculated from Eq. (30) and (8):

$$\varepsilon_0 = 10\,Dq\left(2 - 5\frac{m}{n}\right) \tag{31}$$

By the use of Eq. (20) or (31) an expression for the A_α^q coefficient of the excited state is obtained which allows its calculation from the parameter Dq and R in the same way as shown in the one electron case. For instance, if the ground state Γ_p results from a t_2^n configuration, a transition by excitation of m electrons into an e orbital (yielding a Γ_q state which belongs to the $t_2^{n-m}e^m$ configuration) the perturbation energy of the Γ_q excited state is calculated using Eq. (8), (20) and (30) by

$$\Delta E(\Gamma_q \text{ from } t_2^{n-m}e^m) = A_\alpha^q S_1 = -\frac{m \cdot 50\,Dq}{\sqrt{6}R} S_1 \tag{32}$$

which is with Eq. (21)

$$\Delta E(\Gamma_q) = m\Delta E(e) \tag{33}$$

If Γ_p and Γ_q (excited state) belong to the same electron configuration the potential energy curves are shifted by an equal amount due to vibronic coupling. Therefore no relative shifts are calculated for these states from first order perturbation theory in which configuration interaction (CI) is neglected.

Potential energy curves corresponding to those in Eq. (12) are obtained as well by performing an appropriate coordinate transformation which places the origin of the normal coordinates into the minimum of the ground state potential curve leading to potential minima shifts and stabilization energies which correspond to Eq. (16) and (17) with appropriate vibronic coupling parameters A_α^p, force constants k_α^p and frequencies ω_α^p. All formulas derived for the e $\leftrightarrow$ t_2 single electron case remain valid if parameters of more electron transitions are used. The procedure can be applied also to spinor levels assigned to two-valued representations used for systems with an odd number of electrons and large spin-orbit coupling (vide infra).

When considering configuration interaction (CI) the wave function Ψ_{Γ_p} in Eq. (28) is a CI function

$$\Psi_{\Gamma_p} = \sum_i c_i \Psi_i(\Gamma_p \text{ from } t_2^{n-m}e^m) \tag{34}$$

which combines the functions of symmetry type Γ_p belonging to all electron configurations $t_2^{n-m}e^m$ (m going from 0 to n for $n < 4$ or to 4 for $n \geq 4$). At various occasions more than one level of the same symmetry type Γ_p may result from a certain configuration, in this case the matrix elements of the vibronic perturbation operator with respect to these functions Eq. (28) are equal because they depend only on the electron configuration (cf. Eq. (30)). As an example we may consider the $^2T_1$ states in an octahedral d^3 system. Since this type of multiplet occurs twice in two of the possible configurations, i.e. in t_2^2e and t_2e^2, and not in e^3 [12], we obtain for the CI wave function

$$\Psi(^2T_1) = c_0\Psi_0(t_2^3) + c_1\Psi_1(t_2^2e) + c_1'\Psi_1'(t_2^2e) + c_2\Psi_2(t_2e^2) + c_2'\Psi_2'(t_2e^2) \tag{35}$$

Since Ψ_1 and Ψ_1' (as well as Ψ_2 and Ψ_2') will furnish identical matrix elements of the vibronic coupling operator we can combine the coefficients as $c_1^2 + c_1'^2 = d_1^2$ and $c_2^2 + c_2'^2 = d_2^2$ yielding for the total energy correction Eq. (28) when applying Eq. (30) only on one functions Ψ_m of type Γ_p out of the configuration $t_2^{n-m}e^m$

$$\Delta E(\Gamma_p \text{ of } d^n) = \sum_m d_m^2[(n-m)a_\alpha^t + ma_\alpha^e]S_1 = A_\alpha^p S_1 \tag{36}$$

Any non-diagonal elements $d_m d_l$ must vanish by virtue of orthogonality conditions on t_2 and e orbitals since the coupling operator $\partial V_{LF}/\partial S_1$ is a one-electron operator.

The inclusion of CI therefore results in a much more complex relation between ε_0 and Dq as that given by Eq. (20) and (31). It is obtained from an expression setting Eq. (36) equal to zero in case Γ_p is the ground state of the system and the normal coordinates are measured from its potential minimum. For most of the systems considered the correction due to CI is, however, very small (see below) such that second order perturbation in good approximation can be neglected.

In the many electron case the vertical transition energy from the ground state minimum is not equal to 10 Dq as in Eq. (18) (cf. Fig. 2) but can be calculated from ligand field theory in terms of Dq and Racah parameters B and C. Since this transition corresponds in general to the maximum E_{max} of the absorption or emission band we can use the relation

$$E_{max} = E^{0'} \pm |\Delta E_v| \tag{37}$$

where the plus sign refers to absorption, the minus sign to emission (cf. Fig. 2) and ΔE_v is the stabilization energy of Eq. (17) due to coupling with the mode ν; $E^{0'}$ again is connected to the zero phonon transition E_{0-0} as introduced in Eq. (15).

2.3 Vibronic Coupling Arising from ε and τ_2 Vibrations

Let us now turn to vibronic coupling to degenerate ε and τ_2 vibrations which by linear coupling defined from the Taylor expansion of Eq. (1) supply finite coupling coefficients. We shall, however, consider only systems with non-degenerate initial (ground) states, e.g., $^4A_{2g}$ in octahedral d^3, $^6A_{1g}$ in high spin d^5 or $^1A_{1g}$ in low spin d^6, because others are subject to a Jahn-Teller effect which distorts the nuclear framework to lower symmetry. Molecules with structures different from octahedral symmetry are not within the scope of the present work. Since non-degenerate states do not couple by first order perturbation with degenerate vibrational modes due to symmetry reasons, potential curves of these states will not be shifted along corresponding nuclear coordinates. Therefore we need to investigate only the vibronic coupling and potential energy shifts of excited states, such that the analysis starts from the energy expressions of Eq. (12) in which appropriate notations for the coordinates and parameters describing a $\Gamma_p \rightarrow \Gamma_q$ transition are introduced:

$$
\begin{aligned}
E_q &= E^0 + \tfrac{1}{2} k_v^q Q_i^2 + A_v^q Q_i \\
E_p &= \tfrac{1}{2} k_v^p Q_i^2
\end{aligned}
\qquad (38)
$$

if $i = 2, 3$ refers to a $v = \varepsilon$ and $i = 4, 5, 6$ to a $v = \tau_2$ coupling. The linear coupling coefficient A_v^p for the non-degenerate ground state Γ_p vanishes by symmetry reason.

Since the coupling problem of orbitally degenerate electronic levels Γ_q to ε and τ_2 vibrational modes corresponds to that dealt with in the Jahn-Teller effect we can use the corresponding coupling matrices worked out for this theory[2–4].

2.3.1 Orbital Vibronic Couplings

We start considering vibronic coupling of e and t_2 orbital levels to ε and τ_2 modes proceeding in the same way as in the α-coupling case. In Table 2 the d-orbital perturbation matrices, coordinate shifts and stabilization energies calculated from first order coupling terms are listed for all finite coupling cases of cubic symmetry. An $e \otimes \tau$ coupling vanishes by symmetry reason; the $t \otimes \varepsilon$ and $(e + t) \otimes \tau$ coupling results apply for t_1 as well as t_2 symmetry. The coupling coefficients are calculated from corresponding matrix elements similar to Eq. (8) defined, e.g., for

$$
a_\varepsilon^e = \left\langle \psi_{x^2-y^2} \middle| \left(\frac{\partial V_{LF}}{\partial Q_2}\right)_{Q^0} \middle| \psi_{x^2-y^2} \right\rangle - \left\langle \psi_{z^2} \middle| \left(\frac{\partial V_{LF}}{\partial Q_2}\right)_{Q^0} \middle| \psi_{z^2} \right\rangle \qquad (39)
$$

if the coordinate Q_2 transforms as z^2 [4]. As demonstrated for the α_1 coupling case (Sect. 2.2) the matrix elements are calculated from the d-orbital functions and the potential derivatives of Table 1 yielding

$$
a_\varepsilon^e = +\frac{6}{7}\sqrt{3}\,\frac{Dq}{R}\left(\frac{12}{\eta} + \frac{25}{3}\right) \qquad (40)
$$

In Table 2 the corresponding formulas are listed for all coupling cases considered. The

Table 2. First order vibronic coupling to ligand field d-orbital levels ($k_v = m_v\omega_v^2$)

Coupling	Perturbation matrix, orbital vibronic constants a_v^p	Coordinate shift ΔQ_v [a]	Stabilization energy $\lvert\Delta E_v^p\rvert$
$\gamma_p \otimes \alpha$	$a_\alpha^p Q_1 \cdot \mathit{1}$ [b] for $\gamma_p = \begin{cases} e \quad a_\alpha^e = -\dfrac{1}{\sqrt{6}R}(\varepsilon_0 + 30\,Dq) \\ t_2 \quad a_\alpha^t = -\dfrac{1}{\sqrt{6}R}(\varepsilon_0 - 20\,Dq) \end{cases}$	$\Delta Q_1 = -\dfrac{a_\alpha^p}{k_\alpha}$	$\lvert\Delta E_\alpha^p\rvert = \dfrac{(a_\alpha^p)^2}{2k_\alpha}$
$e \otimes \varepsilon$	$\dfrac{1}{2}a_\varepsilon^e \begin{pmatrix} -Q_2 & Q_3 \\ & Q_2 \end{pmatrix}$ [c] $a_\varepsilon^e = +\dfrac{6}{7}\sqrt{3}\dfrac{Dq}{R}\left(\dfrac{25}{3} + \dfrac{12}{\eta}\right)$	$\Delta Q_\varepsilon^e = \dfrac{a_\varepsilon^e}{2k_\varepsilon}$	$\lvert\Delta E_\varepsilon^e\rvert = \dfrac{(a_\varepsilon^e)^2}{8k_\varepsilon}$
$t \otimes \varepsilon$	$\dfrac{a_\varepsilon^t}{2\sqrt{3}} \begin{pmatrix} Q_2 - \sqrt{3}Q_3 & 0 & 0 \\ & Q_2 + \sqrt{3}Q_3 & 0 \\ & & -2Q_2 \end{pmatrix}$ $a_\varepsilon^t = +\dfrac{12}{7}\dfrac{Dq}{R}\left(\dfrac{25}{3} - \dfrac{9}{\eta}\right)$	$\Delta Q_\varepsilon^t = \dfrac{a_\varepsilon^t}{\sqrt{3}k_\varepsilon}$	$\lvert\Delta E_\varepsilon^t\rvert = \dfrac{(a_\varepsilon^t)^2}{6k_\varepsilon}$
$(e + t) \otimes \tau$	$a_\tau^{et} \begin{pmatrix} 0 & 0 & -\frac{1}{2}Q_6 & -\frac{1}{2}Q_5 & Q_4 \\ & 0 & \frac{\sqrt{3}}{2}Q_6 & -\frac{\sqrt{3}}{2}Q_5 & 0 \\ & & \multicolumn{3}{c}{\dfrac{-a_\tau^t}{\sqrt{6}a_\tau^{et}} \begin{pmatrix} 0 & Q_4 & Q_5 \\ & 0 & Q_6 \\ & & 0 \end{pmatrix}} \end{pmatrix}$ [d] $a_\tau^{et} = -\dfrac{6}{7}\sqrt{3}\dfrac{Dq}{R}\left(\dfrac{5}{3} + \dfrac{4}{\eta}\right)$	ΔQ_τ^{t} [e] $= \dfrac{a_\tau^t}{3k_\tau}$ $a_\tau^t = +\dfrac{12}{7}\sqrt{6}\dfrac{Dq}{R}\left(\dfrac{5}{3} - \dfrac{3}{\eta}\right)$	$\lvert\Delta E_\tau^t\rvert$ [e] $= \dfrac{(a_\tau^t)^2}{9k_\tau}$

[a] For degenerate vibrations v the shifts refer to one of several equivalent minima

[b] Unit matrix of dimension equal to the degeneracy of orbital γ_p

[c] Q_2 (θ) transforms as $z^2 - \frac{1}{2}(x^2 + y^2)$ and Q_3 (ε) as $x^2 - y^2$

[d] Q_4 transforms as xy, Q_5 as xz and Q_6 as yz

[e] Calculated for $t \otimes \tau$ coupling only

parameter η introduced at several other occasions has been used already earlier as low symmetry model parameter in ligand field theory.

In the case of τ_2 coupling the perturbation matrix has to be extended by second order terms since the (e + t) matrix has non-vanishing off-diagonal elements $\langle\psi_e|(\partial V_{LF}/\partial Q_{\tau_2})_{Q^0}|\psi_t\rangle$ intermixing e and t orbitals which become significant for CI calculations between levels belonging to $t_2^{n-m}e^m$ configurations. Bacci noticed[13)] that these "pseudo-Jahn-Teller coupling constants", which we denote by a_τ^{et} in Table 2, are of similar importance for calculating adiabatic potential energy surfaces (APES) of tetrahedral d^1 systems as are 2nd order perturbations due to $e \otimes \varepsilon$ or $t \otimes (\varepsilon + \tau)$ coupling which arise from intermixing with higher levels.

Table 3. Many electron coupling constants for some multiplets resulting from $t_2^{n-m}e^m$ configurations neglecting CI

Number of d electrons	Level	Configuration	A_ε^p	Strong field function
n = 1(5)	$^2T_2$	$t_2(t_2^5)$	$(\pm)a_\varepsilon^t$	a
n = 1(3)	^2E	$e(e^3)$	$(\pm)a_\varepsilon^e$	a
n = 2(4)	$^1T_2$, $^3T_1$	$t_2^2(t_2^4)$	$(\mp)a_\varepsilon^t$	a
	^1E	$t_2^2(t_2^4)$	$(\pm)a_\varepsilon^t$	a
n = 3	$^2E[^1A_1(t_2^2)^2E(e)]$, $^2E[^1E(t_2^2)^2E(e)]$		a_ε^e	b
	$^2T_2[^1T_2(t_2^2)^2E(e)]$, $^2T_1[^3T_1(t_2^2)^2E(e)]$		$-a_\varepsilon^t + \frac{\sqrt{3}}{2} a_\varepsilon^e$	b
	$^2T_1[^1T_2(t_2^2)^2E(e)]$, $^2T_2[^3T_1(t_2^2)^2E(e)]$		$-a_\varepsilon^t - \frac{\sqrt{3}}{2} a_\varepsilon^e$	b
	$^4T_2$	$^3T_1(t_2^2)^2E(e)$	$-a_\varepsilon^t - \frac{\sqrt{3}}{2} a_\varepsilon^e$	b
	$^4T_1$	$^3T_1(t_2^2)^2E(e)$	$-a_\varepsilon^t + \frac{\sqrt{3}}{2} a_\varepsilon^e$	b
n = 6	$^1T_1$, $^3T_1$	t_2^5e	$-a_\varepsilon^t - \frac{\sqrt{3}}{2} a_\varepsilon^e$	c

[a] From Ref. 12
[b] From Ref. 15
[c] Calculated from Clebsch-Gordon- and Wigner-coefficients of Ref. 15

2.3.2 *ε and τ₂ Coupling to Many-Electron States*

The many electron coupling parameters A_ν^p calculated from the ligand field functions Ψ_{Γ_p} can be written, in general, as linear combinations of orbital vibronic constants a_ν^p. These depend on ε_0, Dq, R and in the case of coupling with degenerate modes ε and τ_2 also on η. The dependence on ε_0 which only occurs in connection with α-couplings is removed by using its relation to Dq (see above). The parameter η, however, cannot be determined from octahedral ligand field spectra when only electronic transitions are considered.

The A_ν^p coupling constants are calculated from ligand field parameters, as pointed out in the α-coupling case, at different levels of approximation. Neglecting CI they can be obtained, e.g., from strong field ligand field functions $\Psi_{\Gamma_p}(t_2^{n-m}e^m)$ as in Eq. (28)

$$A_\nu^p = \left\langle \Psi_{\Gamma_p} \middle| \sum_1^n \left(\frac{\partial V_{LF}^{(1)}}{\partial S_\nu}\right)_{S^0} \middle| \Psi_{\Gamma_p} \right\rangle \tag{41}$$

which will be rewritten in one-electron integrals yielding linear combinations of a_ν^e and a_ν^t corresponding to Eq. (30) obtained for α-coupling. In Table 3 ε-coupling coefficients A_ε^p are listed for some of the strong field configurations $t_2^{n-m}e^m$ by giving their relations to the orbital vibronic constants which in turn can be expressed by Dq, η and R using the formulas of Table 2.

For an octahedral d^6 system in an excited t_2^5e configuration we obtain, for example, in case of an ε-coupling to a T_1 ligand field state

$$A_\varepsilon^T = -a_\varepsilon^t - \frac{\sqrt{3}}{2} a_\varepsilon^e = -25 \frac{Dq}{R} \tag{42}$$

We notice that the η-dependence vanishes in this case which is observed occasionally in calculations of other multiplets resulting from various configurations (cf. Table 3). The same equation as Eq. (42) has been derived for d^5 systems also by McClure et al. proceeding in a similar way, when explaining the Jahn-Teller effect in the $^4T_1$ state of Mn^{2+} in cubic environment[14] although the neglect of CI in this case could be rather serious.

In extension of his procedure[11] which allows estimating metal-ligand bond length changes, Hitchman also worked out a formula for Jahn-Teller distortions due to $E \otimes \varepsilon$ couplings by indicating bond changes along the S_2 coordinate (cf. Table 1) in the ε-space that leads to a tetragonal distortion ($\delta_x = \delta_y = -\frac{1}{2}\delta_z$) of

$$\delta_x \approx \frac{n\Delta'}{12\,Rk_\varepsilon} \tag{43}$$

The derivation assumes that the ligand field parameter varies with the inverse power of n and Δ' is the destabilization energy of e orbitals due to the ligand field splitting[16]. The latter parameter is $\Delta' = 10\,Dq$ in the case where metal-ligand π-bonding can be neglected; in general Δ' is, however, an additional parameter which cannot be determined from electronic spectrum without further assumptions. Since this parameter happens to be

$$\Delta' = 3\,e_\sigma\,, \tag{44}$$

e_σ being defined as σ-antibonding parameter in the angular overlap model (AOM), one may use, if possible, results obtained from the AOM for estimating suitable parameter values[17].

Recently the AOM has been extensively used for describing vibronic coupling effects in transition group complexes[13, 18–20]. The orbital vibronic constant a_ε^e, e.g., can be expressed by

$$a_\varepsilon^e = -\sqrt{3}\,\frac{\partial e_\sigma}{\partial R} \tag{45}$$

in which the derivative of e_σ is introduced as additional parameter or must be calculated from further extensive approximations. In octahedral systems the AOM, in general, is less suitable for applications since the number of parameters necessary exceeds that of ligand field theory. For $E \otimes \varepsilon$ and $T \otimes \varepsilon$ coupling the AOM parameters are e_σ, e_π, $\partial e_\sigma/\partial R$ and $\partial e_\pi/\partial R$ while in the present model only Dq and η parameters are necessary also when CI is included. For lower symmetry where level splittings are determined in particular by virtue of the angular parts of the wave functions the application of the AOM by far has more advantages.

The vibronic coupling parameters A_v^q may be also determined for each of the coupling cases more directly from the experiment if the vibrational fine structure is well resolved: the vibrational intervals of the progressions detected in the electronic spectrum will supply the type of coupling which is operative in the system, i.e. α_1, ε and τ_2 vibrational coupling or some combinations of these modes; and from adapting intensity distributions of progressional members the experimental parameters Δ_v are obtained[9] which are connected with the corresponding shifts ΔQ_i of potential energy curves by Eq. (19). Since the effective mass and frequency in this equation also refer to the ground state as do the normal coordinates Q_i, the potential minimum shifts ΔQ_i can be calculated from Δ_v for all coupling cases. The same is true for the stabilization energy of the ground and as well for the excited state by the use of Eq. (17) and again Eq. (19) which both are valid for any of the coupling cases

$$|\Delta E_v| = \tfrac{1}{2}\, k_v \Delta Q_i^2 = \tfrac{1}{2}\, m_v \omega_v^2 \Delta Q_i^2 = \tfrac{1}{2}\, \hbar\omega_v \Delta_v^2 \tag{46}$$

The coupling parameters A_v^q are then calculated from ΔQ_i or ΔE_v by, respectively, Eq. (16) and (17) for coupling with α_1 vibrations and by the corresponding formulas given in Table 2 for the other coupling cases. For $t \otimes \tau$ coupling if t refers to an excited state for which the effective mass is not known, the coupling parameter can only be given in terms of the L_τ coupling constant used in Englman's book[4] which is related to the present coupling parameter by

$$L_\tau = \left(\frac{\hbar}{m_\tau^t \omega_\tau^t}\right)^{1/2} A_\tau^t \tag{47}$$

Introducing the expression for the stabilization energy ΔE_τ^t of Table 2 we obtain

$$L_\tau = 3\sqrt{\hbar\omega_\tau^t / |\Delta E_\tau^t|} \tag{48}$$

containing only the circular frequency which can be determined for the ground as well as for the excited electronic state (see above). The signs of coupling parameters A_v^q and potential minima shifts ΔQ_i, however, cannot be determined from the experiment since they are calculated from the empirical parameters Δ_v which are only obtained by absolute values. The question whether the molecule in the excited state is expanded or compressed compared to the ground state can be answered by applying other theoretical methods[1, 21].

In the case where several distinct vibrational progressions can be resolved in the spectrum, indicating simultaneous coupling with α_1, ε and/or τ_2 modes separate sets of ω_v, β_v and Δ_v parameters are determined for each of the progressions. From these vibronic coupling constants, energy stabilizations and potential minima shifts along different normal coordinates are calculated. The superposition of these effects then supplies the actual distortion of the molecule in the excited state which is manifested by a change of internuclear equilibrium distances. The analysis in terms of ligand field parameters also can be carried out applying the present theory for each normal coordinate separately.

3 Application to Vibronic Spectra

The dynamic ligand field model formulated in the preceding section shall now be used for interpreting ligand field absorption and emission spectra which exhibit distinct vibrational progressions in electronic transitions. But, also in the case when the vibrational fine structure could not be resolved by experiment, the model can be applied when only the intensity distribution of the broad band absorption or emission is used. The additional information which is, however, needed for a spectral analysis would be the knowledge of the type of coupling mode causing the band broadening and the vibrational quantum $\hbar\omega_v$ associated with this mode. Recently it has been shown that this information is available, e.g., from a time resolved photon counting technique[22, 23]. In the present work we will limit ourselves to those cases where ligand field transitions exhibit distinct vibrational progressions for which identification of the type of vibronic coupling is no problem. Also the case where the vibronic spacing in the progression do not match with normal vibrations obtained from corresponding vibrational spectra (IR or Raman), due to the "missing mode effect" (MIME)[24, 25], will not be considered here.

3.1 Vibronic Analysis by Parameter Adaptation

In the literature only relatively few absorption and emission spectra with sufficiently well resolved fine structure have been reported. For most of these a band analysis has been carried out by the authors fitting the parameters Δ_v and occasionally also β_v to the measured intensity distributions in the spectra. In the few cases where they are missing we have determined these data from the published spectra. The results are summarized in Table 4a–c where also stabilization energies, coupling constants and distortion parameters calculated from the spectroscopic parameters are listed. For some of these the error bars must be assumed very large due to the low quality of spectral resolution. The spectra reported for $K_3[Co(CN)_6]$[7, 26, 27] obviously differ appreciably which leads to varying parameter sets also presented in Table 4a.

Some of the parameters obtained from the spectral analysis show characteristic differences which deserve a closer inspection. The small parameter values calculated for the stabilization energy, the coupling constant and the geometry shift of $ReCl_6^{2-}$ and of Cs_2MnF_6 (for the $^2E_g \rightarrow {}^4A_{2g}$ transition in emission), which represent weak coupling cases, are well understood from the d-orbital occupations of the metal ion: in these instances the transitions occur within the t_2^3 configuration while the bands of the other systems (strong coupling cases) are assigned to transitions which undergo a configurational change to $t_2^{n-1}e$ leading to a decrease in the bond strength in the excited state. The large geometry distortion obtained from all spectra reported for $K_3[Co(CN)_6]$ on excitation to $^3T_{1g}$ is probably due to covalent bonding attributed to the Co–CN bond being enhanced by a π back bonding due to the low lying π^* orbitals in CN. An excitation of a π-electron from t_2 to σ antibonding e orbitals then decreases the bond strength more rigorously than usual σ- and π-bonded ligands do compared to NH_3 or halogen compounds. The stabilization energy is correspondingly large, much larger than for the ammonia complex, a fact which also can be explained from bond strength relations expected for these compounds. The lower stabilization energy and coupling constant of

Table 4a. Spectroscopic and vibronic parameters obtained from intensity distributions in the progressions. α-coupling cases

Coupling case	$\Gamma_7, \Gamma_8 \otimes \alpha$	$T \otimes \alpha$				$E \otimes \alpha$
Compound	$[ReCl_6]^{2-}$ doped[9]	$K_3Co(CN)_6$			Cs_2MnF_6[28,29]	Cs_2MnF_6 doped[30,31]
	emission	emission			absorption	emission
Transition	$\Gamma_8(^4A_{2g}) \leftarrow \Gamma_7(^2T_{2g})$	$^1A_{1g} \leftarrow {}^3T_{1g}$			$^4A_{2g} \rightarrow {}^4T_{2g}$	$^4A_{2g} \leftarrow {}^2E$
m_α/amu	35	26			19	19
		7)	26)	27)		
$\hbar\omega_\alpha$/cm^{-1}	346 R[a] [32], E	–	410 E	403 E	500 A	592 E
$\beta = \frac{\omega_{exc}}{\omega_{gr}}$	0.995 f[b]	(1)[c]	1.4 f	(1)	(1)	(1)
Δ_α	0.37 f	5.2 f	4.7 f	4.8 f	≈ 2 fpw	≈ 0.15 fpw
$\lvert\Delta E^{p}_{\alpha}\rvert$/cm^{-1}	24	5500	4720	4450	1000	7
$\lvert A^{p}_{\alpha}\rvert / \frac{erg}{cm} 10^{-4}$	0.5	8.2	6.8	6.6	3.3	0.3
ΔQ_α/pm	2	30	27	27	12	0.8
E_{0-0}[d]/cm^{-1}	–	19500	19200	18700	20400	–

[a] From Raman(R), emission(E), absorption(A)
[b] f fitted (from literature), fpw fitted (present work)
[c] Values in brackets: fixed (not fitted)
[d] Calculated from Eq. (37)

the platinum bromo complex (obtained from the band analysis of the spectrum illustrated in Fig. 3) compared to the chloro complex is also attributed to the degree of metal-ligand bond strength which obviously is weaker in the bromine compound. The geometry distortion indicated by their ΔQ_ε values is calculated, on the other hand, more similar, although the stabilization energies $\Delta E^{E}_{\varepsilon}$ are very different; this is explained from the difference in the excited state potential energy surface which due to the decreased vibra-

Table 4b. ε-coupling cases (notations as in Table 4a)

Coupling case	$E \otimes \varepsilon$			$T \otimes \varepsilon$
Compound	Cs_2PtF_6[33]	K_2PtCl_6[23]	K_2PtBr_6[a]	$RbMnF_3$[14]
	emission			absorption
Transition	$\Gamma_1(^1A_{1g}) \leftarrow \Gamma_3(^3T_{1g})$			$^6A_{1g} \rightarrow {}^4T_{1g}$
m_ε/amu	19	35	80	19
$\hbar\omega_\varepsilon$/cm^{-1}	566 E	310 E	190[b] E, R[32]	470 E
$\beta = \frac{\omega_{exc}}{\omega_{gr}}$	(1)	1 f	0.94 fpw	(1)
Δ_ε	3.5 f	5.4 f	5.1 fpw	2.0 f
$\lvert\Delta E^{p}_{\varepsilon}\rvert$/cm^{-1}	3500	4500	2500	940
$\lvert A^{p}_{\varepsilon}\rvert / \frac{erg}{cm} 10^{-4}$	14	12	8.1	5.2
$\lvert\Delta Q_\varepsilon\rvert$/pm	20	30	24	12
E_{0-0}/cm^{-1} Eq. (37)	18350	19500	16100	18330
E_{0-0}/cm^{-1} exp	18250[33]	–	–	18300[14]

[a] Own measurement, cf. Fig. 3
[b] Anharmonicity observed

Table 4c. $(\alpha + \varepsilon)$-coupling cases (notations as in Table 4a)

Coupling case	$T \otimes (\alpha + \varepsilon)$							
Compound	$[CrCl_6]^{3-}$ doped[34, 35] emission		$[Cr(NH_3)_6]^{3+}$ 10) a absorption		$[Co(NH_3)_6]^{3+}$ 36) a 37) b absorption		$[Rh(NH_3)_6]^{3+}$ 38) emission	
Transition	$^4A_{2g} \leftarrow {}^4T_{2g}$		$^4A_{2g} \rightarrow {}^4T_{2g}$		$^1A_{1g} \rightarrow {}^1T_{1g}$		$^1A_{1g} \leftarrow {}^3T_{1g}$	
m/amu	35		≈17		≈17		≈17	
	α	ε	α	ε	α	ε	α	ε
$\hbar\omega/cm^{-1}$	300 E	240 E	462 R	412 R	490 R	440 R	505 R	482 R
$\beta_v = \frac{\omega_{exc}}{\omega_{gr}}$	(1)	(1)	(1)	(1)	(1)	(1)	(1)	(1)
Δ_v	2.7 f	2.2 f	2.2 f	2.1 f	2.7 f	2.4 f	0.8 f	3.7 f
$\lvert\Delta E_v^p\rvert/cm^{-1}$	1100	600	1100	910	1800	1300	160	3300
$\lvert A_v^p\rvert / \frac{erg}{cm}\, 10^{-4}$	2.83	2.89	3.1	4.3	4.1	5.4	1.3	9.5
$\Delta Q_v/pm$	15	14	14	15	18	16	5	24
E_{0-0}/cm^{-1} Eq. (37)	≈12 000[c]		20 200		18 800		20 460	
E_{0-0}/cm^{-1} exp.	11 890[34]		19 960[10]		19 010[36]		≥20 500[38]	

[a] As $(ClO_4)_2Cl \cdot KCl$ double salt; parameters given refer to this compound
[b] As $[Ir(CN)_6]^{3-}$ salt; band analysis yields Δ values ≈10% larger
[c] E_{max} uncertain due to large vibrational fine structure

tional quanta of the bromo complex has a smaller gradient as that of the chloro complex. The distinct frequency effect, i.e. the deviation of β from unity, in the $PtBr_6^{2-}$ complex may add to the excited state distortion of this molecule which is relatively large compared to what is expected from the small stabilization energy. The only spectrum which by inspection exhibits two well resolved progressions is that of $CrCl_6^{3-}$ [34]. The spectra of the other systems in which a combined coupling with α_1- and ε-modes is operative (cf. Table 4c) were analyzed by different procedures using spectroscopic data additional to the intensity distribution in the progression[10, 36, 38].

3.2 Parameter Calculations by the Dynamic Ligand Field Model

We now will use the dynamic ligand field model for calculating the vibronic coupling constants from the ligand field parameters and compare the results with the data which are obtained by fitting the spectral intensities of the spectra measured in absorption or emission (Table 4).

3.2.1 α-Coupling Cases

The applications will focus in particular on the totally symmetric coupling cases since for these the corresponding parameters can be calculated from any configuration using only the cubic ligand field parameter Dq and the metal-ligand bond length R (Sect. 2.2). The best results will be expected from a perturbation calculation with high quality ligand field functions, i.e. which include configuration interaction (CI) and, if necessary, spin-orbit coupling. However, since in the α-coupling case many off-diagonal matrix elements

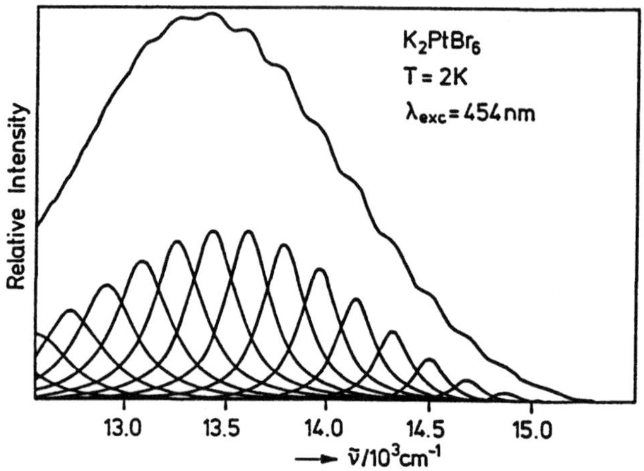

Fig. 3. Low temperature emission spectrum of a K_2PtBr_6 powder sample by Laser excitation at 454 nm wavelength (own measurement) and vibrational analysis by Lorentzian band distributions

between wave functions belonging to different electron configurations vanish by symmetry reason or are small due to the choice of basis functions in the strong field coupling scheme, the calculation can be limited by neglection of CI. The computational effort is certainly smaller than that for coupling with degenerate vibrational modes. The results calculated with and without CI which are obtained for some representative α-coupling systems mentioned above are summarized in Table 5. For reasons of comparison the α-coupling parameters of K_2PtCl_6 which only exhibits ε vibronic coupling have been included as well. In the table we also give the ligand field parameters Dq, Racah parameters B and C, spin-orbit coupling parameters ζ[5,6] and the metal-ligand distances R[39] from which the coupling constants are calculated. Some of these data are known only approximately which affects, of course, the quality of the calculated results.

Table 5. Parameter calculations from the dynamic ligand field model for vibronic coupling with the totally symmetric vibrational mode α_1, *A* without ($\varepsilon_0 = 20\,Dq$) and *B* with configuration interaction (CI) considering full d^n matrices (all energies in cm^{-1})

Compound		$[ReCl_6]^{2-}$	$K_3Co(CN)_6$	$[CrCl_6]^{3-}$	$[Co(NH_3)_6]^{3+}$	K_2PtCl_6
Dq		3000	3400	1300	2400	2840
B		334	440	570	660	260
C		1872	1760	3420	2640	1790
ζ		2140	500	170	500	3400
R/pm		230[a]	189[b]	230[a]	200[a]	233[b]
A	$A^p_\alpha / \frac{erg}{cm}$	0	$-7.3 \cdot 10^{-4}$	$-2.28 \cdot 10^{-4}$	$-4.9 \cdot 10^{-4}$	$-4.9 \cdot 10^{-4}$
B	ε_0	19.8 Dq	19.9 Dq	≈ 20 Dq	19.7 Dq	19.2 Dq
	$A^p_\alpha / \frac{erg}{cm}$	$-1.6 \cdot 10^{-5}$	$-7.3 \cdot 10^{-4}$	$-2.28 \cdot 10^{-4}$	$-4.8 \cdot 10^{-4}$	$-4.8 \cdot 10^{-4}$
	$\lvert E^p_\alpha \rvert$	3	5460	710	3180	290
	ΔQ^p_α/pm	0.7	24	12	26	8

[a] Estimated from ionic radii
[b] Ref. 39

Obviously the inclusion of CI does not have much influence on the results such that for the present systems the calculations considering only the main electron configurations involved will be adequate. In the weakly coupled case of $ReCl_6^{2-}$ the only contributions to the vibronic coupling parameter arise from CI terms.

The comparison with the α-coupling parameters obtained from the measured spectra (Table 4) shows a satisfactory agreement for the strongly coupled systems. In the case of the weakly coupled system, i.e. $ReCl_6^{2-}$, the result is off by a factor of 3. However, at least the order of magnitude (10^{-5} erg/cm) is well reproduced, the corresponding parameter values for the strongly coupled cases (10^{-4} erg/cm) being one decimal place higher. The larger discrepancies noticed in the stabilization energies ΔE_α^p are due to the fact that the error in the coupling constants propagates more significantly into this parameter compared to others since ΔE_α^p is a squared function of A_α^p (cf. Table 2).

The α-parameters obtained for K_2PtCl_6 explain why in the emission spectrum only a progression in ε vibrational intervals is observed: the small stabilization energy of 290 cm^{-1} calculated for α_1 coupling (Table 5) cannot compete with the stabilization arising from ε coupling, i.e. 4500 cm^{-1} (Table 4b); consequently the emission usually originating from the lowest excited state will occur from ε potential minima instead of starting from the α_1 minimum.

3.2.2 ε-Coupling Cases

Finally we consider some of the systems more closely where ε coupling is operative. Vibrational progressions originating from τ_2 vibronic coupling have not been found so far in optical spectra of transition metal compounds. Since orbital vibronic constants a_ε^e and a_ε^t are functions of the parameters Dq, η and R (cf. Table 2) the multiplet coupling constants A_ε^p will in general also depend on these parameters. In some cases when CI is neglected certain transition energies depend only on Dq and R: e.g. for some d^3 and d^6 systems listed in Table 4c the first order coupling parameter of the excited states $^4T_{2g}(t_2^2e)$ and $^1T_{1g}$ or $^3T_{1g}(t_2^5e)$ is $A_\varepsilon^T = -25$ Dq/R (cf. Eq. (42) and Table 3). The corresponding α parameters are calculated from Eq. (32) which is valid for any symmetry type Γ_q of the excited state $t_2^{n-m}e^m$ configuration, i.e. $A_\alpha^T = -50\ Dq/\sqrt{6}\,R$ for m = 1. Considering the relation

$$\frac{A_\alpha^T}{A_\varepsilon^T} = \frac{2}{\sqrt{6}} = 0.816 \tag{49}$$

we obtain a constant which compares well with the corresponding values calculated from the coupling parameters of Table 4c determined from adapting intensity distributions of measured progressions. The discrepancy observed for the Rh compound indicates an erroneous interpretation of this spectrum. Already the large differences reported for the parameter values which have been obtained from the band analysis cannot be reconciled with the spectrum: the high ε stabilization energy of 3300 cm^{-1} compared to an α stabilization of only 160 cm^{-1} would only explain a progression in ε vibrational quanta, an α progression being very weak and probably too short to be detected.

Relative stabilization parameters of T states are calculated using corresponding expressions from Table 2. With the force constants $k_v = m_v\omega_v^2$ we obtain

Table 6. Relative coupling parameters and stabilization constants of T states obtained from the experiment and calculated in first order from the dynamic ligand field theory. For the discrepancies noticed in case of the Rh compound see text

		$[CrCl_6]^{3-}$	$[Cr(NH_3)_6]^{3+}$	$[Co(NH_3)_6]^{3+}$	$[Rh(NH_3)_6]^{3+}$
$A_\alpha^T/A_\varepsilon^T$	exp[a]	0.98	0.72	0.76	0.14
	calc[b]	←	——— 0.816	———	→
$\Delta E_\alpha^T/\Delta E_\varepsilon^T$	exp[a]	1.83	1.21	1.38	0.05
	calc[c]	1.28	1.6	1.6	1.8

[a] From Table 4c
[b] Eq. (49)
[c] From Eq. (51) using the fundamentals $\hbar\omega_v$ of Table 4c

$$\frac{\Delta E_\alpha^T}{\Delta E_\varepsilon^T} = 3\frac{(A_\alpha^T)^2}{(A_\varepsilon^T)^2}\cdot\frac{k_\varepsilon}{k_\alpha} = 3\left(\frac{A_\alpha^T}{A_\varepsilon^T}\right)^2\cdot\left(\frac{\omega_\varepsilon}{\omega_\alpha}\right)^2 \tag{50}$$

in which the masses are $m_\alpha = m_\varepsilon$. With Eq. (49) we get in first order (neglecting CI)

$$\frac{\Delta E_\alpha^T}{\Delta E_\varepsilon^T} = 2\left(\frac{\omega_\varepsilon}{\omega_\alpha}\right)^2 \tag{51}$$

This relation, calculated from the fundamentals $\hbar\omega_v$ listed in Table 4c, may be also compared with the experimental data (cf. Table 6) resulting from adaptation to the measured spectrum. The agreement is not as satisfactory as that obtained for the preceding results which again may be explained by the larger error propagation in ΔE_v^p due to its quadratic functional dependence on A_v^p or, respectively, in Eq. (50) on ω_v. In addition, the influence of CI on the degree of ε coupling could be the reason for some of the differences. Calculations of these coupling constants including CI are, however, rather elaborate since other than in the α-coupling case many off-diagonal elements arising between CI functions have to be considered. An estimation on how much these interactions affect the vibronic coupling parameters is not easy to be obtained. Since the calculated formulas will depend also on the second ligand field parameter η which cannot be determined from electronic transitions of octahedral systems using common ligand field theory, a comparison of theory and experiment will not be possible. The expressions can, however, be used for a calculation of η values from Dq, R and the experimental parameter Δ_ε of Eq. (19), which is obtained from adaptation to the intensity distribution of the vibrational fine structure, applying relevant formulas for the excited state distortions ΔQ_ε^p and coupling constants A_ε^p derived in the preceding sections.

4 Conclusion

The extension of ligand field theory including the dynamics of the ligands in complex molecules allows for calculating vibronic coupling constants from common ligand field parameters. The results can be compared with data obtained by fitting relative intensities of vibrationally resolved spectra to theoretical band profile functions. Although in the present work only octahedral systems have been considered the model will be applicable also to compounds of lower symmetry.

For an evaluation of the results, it should be remembered that the dynamic model depends on ligand field derivatives being less well defined than the ligand field potentials themselves; the results therefore cannot be expected on the same level of quality as those obtained from standard ligand field theory which is able to supply only data on electronic states. Also the parameters determined from fitting intensities of vibrational fine structures are subject to large error limits which is particularly true for poorly resolved spectra of superimposed progressions.

Acknowledgment. We appreciate the financial support of the Fonds der Chemischen Industrie, Frankfurt/Main.

5 References

1. Ballhausen, C. J.: Molecular electronic structures of transition metal complexes. McGraw-Hill, New York (1979)
2. Bersuker, I. B.: The Jahn-Teller effect and vibronic interaction in modern chemistry. Plenum, New York (1984)
3. Perlin, Y. E., Wagner, M. (eds.): The dynamical Jahn-Teller effect in localized systems. North Holland, Amsterdam (Modern problems in condensed matter sciences, vol. 7) (1984)
4. Englman, R.: Jahn-Teller effect in molecules and crystals. Wiley Interscience, New York (1972)
5. Ballhausen, C. J.: Introduction to ligand field theory. McGraw Hill, New York (1962)
6. Schläfer, H. L., Gliemann, G.: Einführung in die Ligandenfeldtheorie. Akademische Verlagsgesellschaft, Frankfurt/Main (1980)
7. Viaene, L., D'Olieslager, J., Ceulemans, A., Vanquickenborne, L. G.: J. Amer. Chem. Soc. *101,* 1405 (1979)
8. Van Vleck, J. H.: J. Chem. Phys. *7,* 72 (1939)
9. Kupka, H., Enßlin, W., Wernicke, R., Schmidtke H.-H.: Mol. Phys. *37,* 1693 (1979)
10. Wilson, R. B., Solomon, E. I.: Inorg. Chem. *17,* 1729 (1978)
11. Hitchman, M. A.: Inorg. Chem. *21,* 821 (1982)
12. Griffith, J. S.: The theory of transition metal ions. Cambridge University Press, Cambridge (1961)
13. Agresti, A., Ammeter, J. H., Bacci, M.: J. Chem. Phys. *81,* 1861 (1984)
14. Chen M. Y., McClure, D. S., Solomon, E. I.: Phys. Rev. B *6,* 1690, 1697 (1972)
15. Sugano, S., Tanabe, Y., Kamimura, H.: Multiplets of transition metal ions in crystals. Academic, New York (1970)
16. Deeth, R. J., Hitchman, M. A.: Inorg. Chem. *25,* 1225 (1986)
17. Smith, D. W.: Structure and Bonding *35,* 87 (1978)
18. Bacci, M.: Chem. Phys. *40,* 237 (1979)
19. Bacci, M.: Structure and Bonding *55,* 67 (1983)
20. Bill, H.: Observation of the Jahn-Teller effect with electron paramagnetic resonance, in Ref. 3
21. Hakamata, K., Urushiyama, A., Degen, J., Kupka, H., Schmidtke, H.-H.: Inorg. Chem. *22,* 3519 (1983)
22. Degen, J., Schmidtke, H.-H., Chatzidimitriou-Dreismann, C. A.: Theoret. Chim. Acta *67,* 37 (1985)
23. Schmidtke, H.-H.: in: Lever A.B.P. (ed.) Excited states and reactive intermediates, American Chemical Society, Washington (ACS symposium series no. 307) (1986)
24. Tutt, L., Tannor, D., Heller, E. J., Zink, J. I.: Inorg. Chem. *21,* 3858 (1982)
25. Tutt, L., Zink, J. I., Heller, E. J.: Inorg. Chem. *26,* 2158 (1987)
26. Mazur, U., Hipp, K. W.: J. Phys. Chem. *84,* 194 (1980)

27. Oelkrug, D., Radjaipour, M., Eitel, E.: Spectrochim. Acta A *35,* 167 (1979)
28. Novotny, D. S., Sturgeon, G. D.: Inorg. Nucl. Chem. Lett. *6,* 455 (1970)
29. Reisfeld, M. J., Matwiyoff, N. A., Asprey, L. B.: J. Mol. Spectr. *39,* 8 (1971)
30. Helmholz, L., Russo, M. E.: J. Chem. Phys. *59,* 5455 (1973)
31. Chodos, S. L., Black, A. M., Flint, C. D.: J. Chem. Phys. *65,* 4816 (1976)
32. Nakamoto, K.: Infrared spectra of inorganic coordination compounds, 3rd edition. Wiley Interscience, New York (1978)
33. Laurent, M. P., Patterson, H. H., Pike, W., Engstrom, H.: Inorg. Chem. *20,* 372 (1981)
34. Güdel, H. U., Snellgrove, T. R.: Inorg. Chem. *17,* 1617 (1978)
35. Güdel, H. U.: in: Lever A.B.P. (ed.) Excited states and reactive intermediates, American Chemical Society, Washington (ACS symposium series no. 307) (1986)
36. Wilson, R. B., Solomon, E. I.: J. Amer. Chem. Soc. *102,* 4085 (1980)
37. Komi, Y., Urushiyama, A.: Bull. Chem. Soc. Jpn. *53,* 980 (1980)
38. Hakamata, K., Urushiyama, A., Kupka, H.: J. Phys. Chem. *85,* 1983 (1981)
39. Wyckhoff, R. W.: Crystal structures, Vol. III, 2nd edition. Wiley Interscience, New York (1981)

The Epikernel Principle

A. Ceulemans and L. G. Vanquickenborne

Department of Chemistry, University of Leuven, Celestijnenlaan 200F, B-3030 Leuven, Belgium

An epikernel is an intermediate subgroup in the decomposition scheme of a given point group. The epikernel principle states that the preferred distortions of Jahn-Teller unstable molecules are directed towards the maximal allowed epikernels of the undistorted parent group. The group theoretical foundations of this principle are explained, and a wide variety of applications in different areas of chemistry is discussed.

1 Introduction . . . 126

2 The Symmetry of Displacement Coordinates . . . 126

3 Kernels and Epikernels . . . 128
- 3.1 The Coordinate Representation is Non-Degenerate . . . 128
- 3.2 The Coordinate Representation is Degenerate . . . 129
- 3.3 The Coordinate Representation is Reducible . . . 134

4 The Epikernel Principle . . . 136

5 JT Surfaces for Electronic Doublets and Triplets . . . 139
- 5.1 Electronic Doublets . . . 139
  - 5.1.1 The Cubic $E \times e$ Problem . . . 139
  - 5.1.2 The Tetragonal $E \times (b_1 + b_2)$ Problem . . . 141
  - 5.1.3 Generalization . . . 142
- 5.2 Electronic Triplets . . . 144
  - 5.2.1 The Icosahedral $T \times h$ Problem . . . 145
  - 5.2.2 The Cubic $T \times (e + t_2)$ Problem . . . 145

6 Applications . . . 149
- 6.1 Organic Radicals . . . 150
- 6.2 Organic Diradicals . . . 151
- 6.3 Transition-Metal Complexes . . . 152
- 6.4 Cluster Compounds . . . 153

7 Conclusions . . . 156

8 References and Notes . . . 157

© Springer-Verlag Berlin Heidelberg 1989

1 Introduction

The Jahn-Teller (JT) theorem asserts the existence of spontaneous symmetry breaking distortions in degenerate electronic states of non-linear molecules [1, 2]. While this definition emphasizes the symmetry destroying nature of JT instabilities, it tends to obscure the simultaneous importance of symmetry conservation during the JT distortion process. The latter aspect is incorporated in the *epikernel principle* [3]. According to this principle, Jahn-Teller distortions, which lead to stable minima on the potential energy surface near an electronic degeneracy, preserve larger symmetry groups, than could be expected from an unrestricted action of the JT distorting forces.

Previously we presented a detailed discussion of both antithetical aspects of the JT effect, on the basis of a case study of a tetrahedral instability [3]. The present article provides a generalized treatment, which focuses on the common structure of all types of JT instabilities. The treatment is followed by an exhaustive overview of the applications of the epikernel principle in all point group symmetries, except for the fourfold and fivefold degenerate representations in icosahedral symmetry. The purely mathematical aspects of this treatment have been published in a separate paper [4].

In this survey, we will not be concerned with the dynamic aspects of the JT theorem. In fact the concepts of symmetry destruction and symmetry conservation, to which the epikernel principle refers, are properties of the adiabatic potential energy surface only. When the kinetic energy of the nuclei is included, the JT coupling does not destroy the initial symmetry, but just replaces the electronic degeneracy by a vibronic degeneracy, due to dynamic tunneling between the different equivalent minima on the JT surface.

2 The Symmetry of Displacement Coordinates

Before proceeding with an exposition of the epikernel principle, it is well to stress that our approach is not limited to structural predictions. More generally, we are interested in the global properties of potential surfaces and the possibilities of chemical (isomerization or exchange) reactions as specific paths on these surfaces. Obviously, in the neighborhood of a degeneracy, the Jahn-Teller theorem and the epikernel principle will strongly influence the structure of the surfaces and the orientation of the preferred reaction paths.

First, we have to make a general remark on the symmetry of nuclear displacement coordinates. In the chemical literature there are two alternative ways of characterizing the symmetry of a distortion coordinate [5, 6]. One way is by specifying *the group of symmetry elements, which are conserved during the displacements of the nuclei.* This group will always be a subgroup of the point group at the starting configuration, from which the displacement is activated. The other way is by means of *irreducible representations of the parent group:* this method describes the transformational behavior of the reaction coordinates under the symmetry operations of the (high symmetry) reference configuration.

The former method is exemplified by the Woodward-Hoffmann approach, which emphasizes the conservation of symmetry along a reaction path [7]. Clearly, this approach assigns reaction pathways by means of subgroups. Hence, as an example, the

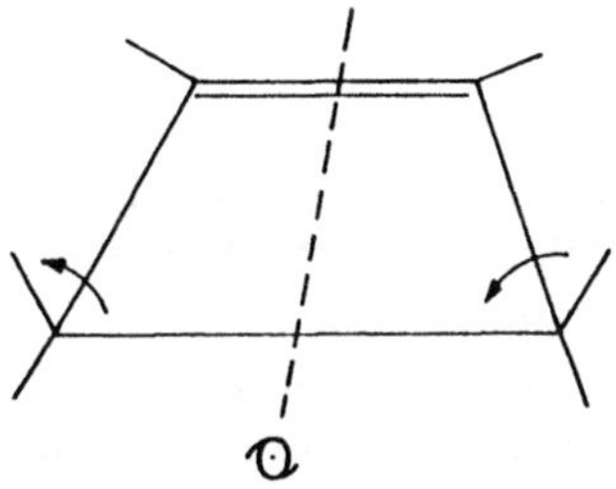

thermal ring opening of cyclobutene, is shown to require a conrotatory mode. This pathway is characterized as the C_2-mode, the twofold axis being the only symmetry element of the starting C_{2v} group which is conserved during the reaction.

The alternative method of specifying distortion coordinates is used in the application of the JT effect. The rule of the JT instability states [8] that the JT active distortion coordinates of a degenerate electronic state of symmetry type Γ must span representations Λ belonging to the non-totally-symmetric part of the symmetrized direct product of Γ.

$$\Lambda \in [\Gamma \times \Gamma] - A_1 \tag{1}$$

Clearly this rule assigns coordinate symmetries by means of irreducible representations, instead of conserved subgroups. A similar methodology is used in the work of Pearson [9] and Bader [10].

It is obvious that the labelling methods used in the Woodward-Hoffmann and Jahn-Teller approaches must somehow be related. This is particularly evident in the cases where only one mode of a single symmetry type has to be considered, as in the example of cyclobutene. Indeed the conrotatory displacement is symmetric under the C_2 axis, which is conserved, and antisymmetric with respect to the symmetry planes, which are destroyed. Hence it will transform as the a_2 representation and we obtain the following equivalence relation:

$$C_{2v} : C_2 \leftrightarrow a_2 \tag{2}$$

Such a simple one-to-one correspondence will always be found when the distortional mode spans a non-degenerate irreducible representation. A general method to generate these representations for Woodward-Hoffmann type reactions has been designed by Halevi and is known as Orbital Correspondence Analysis in Maximum Symmetry (OCAMS) [11, 12].

If the representations of the displacement coordinates are degenerate or reducible, i.e. if there are *several* symmetry components involved, a more intricate relationship between both ways of labeling can be expected. Indeed in these cases one is no longer dealing with a single distortion coordinate, but with a distortion space, covering several dimensions. In principle different directions in this space may have different symmetries and thus may conserve different subgroups. This is where the concepts of kernel and epikernel [13, 14] become useful, as will be discussed in the next paragraph.

3 Kernels and Epikernels

A symmetry operation, which leaves a distortion coordinate unchanged, will also act trivially upon the resulting distorted nuclear configuration. It thus belongs to the group of congruence operations of all structures along the distortion coordinate and therefore is said to be *conserved* during the nuclear motion. In contrast, an operation, which changes the distortion coordinate, does not carry the distorted configuration into itself, and thus will be lost as soon as the distortion starts off. Consequently *a symmetry operation will be conserved during a nuclear displacement if and only if it leaves the displacement coordinate invariant.* This simple principle is at the basis of the relationship between mode representations and conserved subgroups. When applying this principle in practice three different cases must be distinguished, depending on the nature of the coordinate representation.

3.1 The Coordinate Representation is Non-Degenerate

In this section we will only deal with real non-degenerate representations, i.e. representations with +1 and −1 characters. Unidimensional representations with complex characters always occur in conjugate pairs and hence will be discussed in the section on reducible representations (Sect. 3.3). As far as the real non-degenerate representations are concerned, the application of the symmetry conservation principle is straightforward. The conserved subgroup is the subgroup of all symmetry elements with +1 entry in the character table for that representation. This subgroup will be called the kernel of the coordinate representation [13]. It is symbolized as K(G, Λ), where G is the parent group and Λ the representation. Taking the example of the ring opening in cyclobutene, the character table for the a_2 mode reads:

C_{2v}	E	C_2	σ_1	σ_2
a_2	+1	+1	−1	−1

The mirror planes with −1 characters do not leave the a_2 coordinate invariant. One thus retains the C_2 subgroup as the kernel of the conrotatory path. Equation (2) may therefore be rewritten as:

$$C_2 = K(C_{2v}, a_2) \tag{3}$$

A few representative examples of kernels have been listed in Table 1. More complete tables are available in the literature [13, 15]. The paper by Murray-Rust et al. [6] provides a comprehensive description of some interesting properties of kernel groups.

Although symmetry operations with −1 characters do not survive symmetry lowering distortions, they play an important role in the analysis of distortion coordinates. Indeed when acting upon a given coordinate, these elements will reverse the direction of the displacement. In this way distorted configurations at coordinates Q and −Q are strictly equivalent to each other. As an example we may consider the b_{2g} bending mode of a square planar molecule. This mode yields a rectangular geometry with D_{2h} symmetry.

Table 1. Kernel subgroups of one-dimensional representations[a]

G	Λ	K(G, Λ)	G	Λ	K(G, Λ)
C_{2v}	a_1	C_{2v}	D_{4h}	a_{1g}	D_{4h}
	a_2	C_2		a_{2g}	C_{4h}
	b_1	$C_s(\sigma_v)$		b_{1g}	$D_{2h}(C_2')$
	b_2	$C_s(\sigma_v')$		b_{2g}	$D_{2h}(C_2'')$
				a_{1u}	D_4
D_{3d}	a_{1g}	D_{3d}		a_{2u}	C_{4v}
	a_{2g}	S_6		b_{1u}	$D_{2d}(C_2')$
	a_{1u}	D_3		b_{2u}	$D_{2d}(C_2'')$
	a_{2u}	C_{3v}			
			O_h	a_{1g}	O_h
				a_{2g}	T_h
				a_{1u}	O
				a_{2u}	T_d

[a] G is the parent group, Λ an irreducible representation and K(G, Λ) the corresponding kernel. If necessary, equivalent kernels are distinguished by their principal symmetry elements.

The in-plane C_2 axes which are conserved during this bending are labeled C_2''. Accordingly the D_{2h} kernel is further specified as $D_{2h}(C_2'')$ (Cf. Table 1 and Eq. 4). The $Q_{b_{2g}}$ and

$$D_{2h}(C_2'') = K(D_{4h}, b_{2g}) \tag{4}$$

$-Q_{b_{2g}}$ paths may be interchanged by various symmetry operations of the starting configuration, such as for instance the C_4 axis.

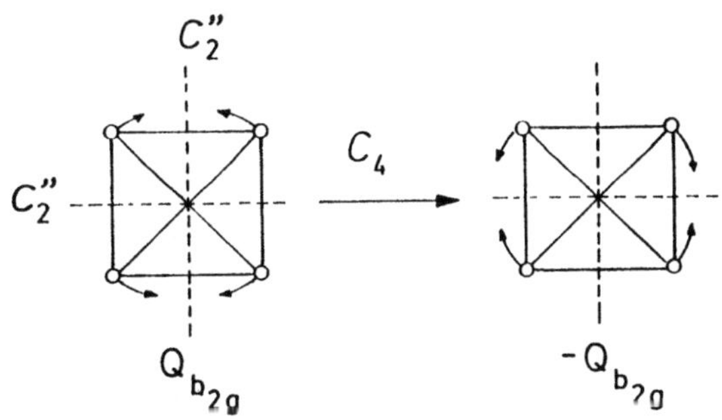

The resulting distorted configurations are clearly equivalent. Totally symmetric modes cannot show this kind of equivalence, since there are no symmetry operations that can reverse their sense.

3.2 The Coordinate Representation is Degenerate

A degenerate representation describes the symmetry of a set of coordinates. The elements of this set are called the *components* of the representation. The components span a *distortion space.* As we have already indicated, the relationship between the symmetry of this space and its conserved subgroups is more intricate than in the case of a non-degenerate representation, because the degeneracy leaves the direction of the distortion unspecified. In order to draw a map of all conserved subgroups, one thus has to scan all directions of the distortion space. Nonetheless there is a minimal group of symmetry elements, which must be conserved in any case, no matter which spatial direction is chosen. This group consists of those symmetry operations, that leave *all* allowed distortions invariant. The effect of these operations on the distortion space is described by the unit matrix. In analogy with the previous definition, this minimal subgroup is said to form the *kernel* of the representation. As in the case of a non-degenerate representation, the kernel of a degenerate mode may immediately be obtained from the character tables by collecting all elements with the same character as the identity operation. As an example the kernel of the degenerate e_g representation in O_h is verified to be D_{2h}, being the subgroup of all elements with +2 characters.

O_h	E	$8C_3$	$6C_4$	$3C_4^2$	$6C_2$	i	$8S_6$	$6S_4$	$3\sigma_h$	$6\sigma_d$
e_g	2	−1	0	2	0	2	−1	0	2	0

More specifically this subgroup may be labeled $D_{2h}(C_4^2)$, since its twofold axes correspond to the three C_4^2 axes of the parent O_h group. Hence one has:

$$D_{2h}(C_4^2) = K(O_h, e_g) \tag{5}$$

This symmetry lowering from O_h to D_{2h} is exemplified by one of the two stretchings of the e_g type.

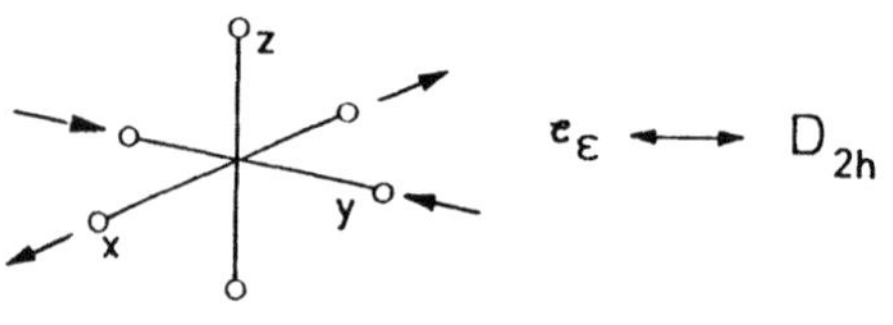

The component shown is the e_ε partner, transforming as $x^2 - y^2$. Most other distortions in the e_g space will also only conserve the symmetry elements of the D_{2h} kernel. However there are a few privileged directions, which conserve more symmetry elements than present in the kernel. One of these distortions is the e_θ component transforming as $2z^2 - x^2 - y^2$. This distortion is seen to preserve a tetragonal symmetry around the z axis. Its conserved subgroup will be denoted as D_{4h}^z.

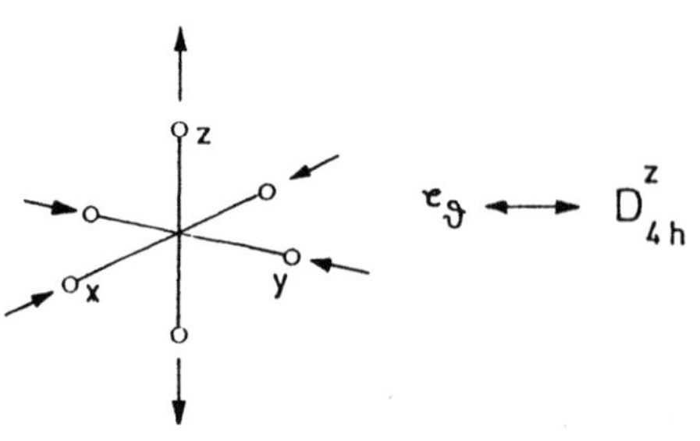

Such a subgroup, which is only conserved in a part of the distortion space, is termed an *epikernel* [14] of Λ in G, and symbolized as E(G, Λ). Hence:

$$D_{4h} = E(O_h, e_g) \tag{6}$$

Clearly epikernels are intermediate subgroups between the parent group (which exists only in the origin of the space), and the kernel group, which is conserved in all distorted structures. This may be expressed as follows:

$$G \supset E(G, \Lambda) \supset K(G, \Lambda) \tag{7}$$

Component choices, such as e_θ, which conserve epikernels, often also give rise to special standard forms of the corresponding representation matrices, and therefore coincide with the so-called *canonical* components.

Since the three principal axes of an octahedron are equivalent, the e_g space must contain two more asymmetric stretchings which preserve the same tetragonal symmetry around the x and y axis. These directions are described by the following coordinates:

$$D^x_{4h}: \frac{\sqrt{3}}{2}Q_\varepsilon - \frac{1}{2}Q_\theta$$

$$D^y_{4h}: -\frac{\sqrt{3}}{2}Q_\varepsilon - \frac{1}{2}Q_\theta \tag{8}$$

where Q_ε and Q_θ denote the distortion coordinates, transforming as e_ε and e_θ respectively. A further search of the e_g space yields no other epikernels.

Figure 1 represents the spatial distribution of the epikernels in e_g space. The figure completely specifies the desired relationship between the e_g mode and its conserved subgroups.

The center of the figure corresponds to the octahedral parent configuration. Tetragonal configurations form three distortional directions, lying at 120° from each other and indicated as D^z_{4h}, D^x_{4h}, D^y_{4h} (see also Eq. 8). The points between these axes represent distorted structures with D_{2h} kernel symmetry only. As the figure illustrates, the spatial distribution of conserved subgroups has a formal trigonal symmetry: for each distorted D^z_{4h} structure there are *two* more equivalent epikernel structures along the D^x_{4h} and D^y_{4h} directions, while any arbitrary kernel point is repeated *five* more times. However, distortions in opposite directions in general do not lead to equivalent configurations, contrary to the case of the b_{2g} mode in the previous section. This is because there are no symmetry elements in O_h, for which the e_g representation is totally antisymmetric.

The number of equivalent points in Fig. 1 illustrates an interesting group theoretical property, which follows from Lagrange's theorem on the orders of subgroups [16]. According to this property the number of equivalent epikernel or kernel directions in a given distortion space, n_E or n_K, equals the quotient of the respective group orders, as expressed by the following equations:

$$n_E = |G|/|E(G, \Lambda)| \tag{9}$$

$$n_K = |G|/|K(G, \Lambda)| \tag{10}$$

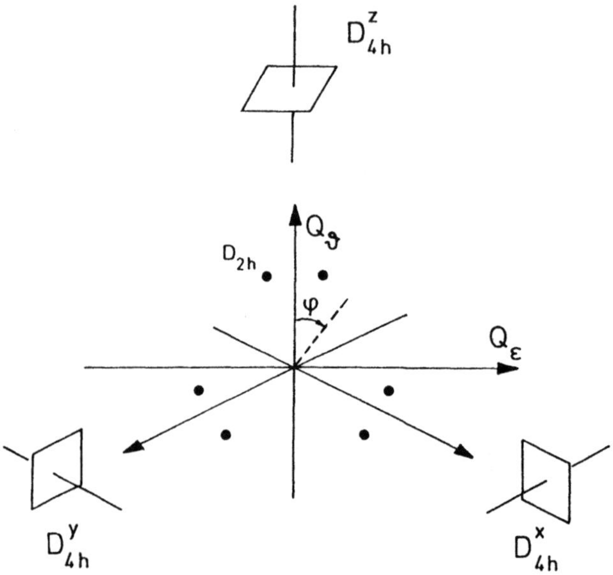

Fig. 1. Schematic representation of the distortion space for the e_g stretching mode in an octahedral molecule. The coordinate axes are the Q_θ and Q_ε distortions, as specified in the text. The angle φ defines the direction of distortion. The center of the figure refers to the parent O_h structure. All points on the Q_θ axis correspond to configurations with D^z_{4h} epikernel symmetry. Similarly, the axes at φ = 120° and φ = 240° represent configurations with resp. D^x_{4h} and D^y_{4h} epikernel symmetries. All other points refer to distorted structures, that only conserve the D_{2h} kernel symmetry. For any arbitrary kernel point, the number of equivalent distorted configurations is six – as illustrated by the *black circles*

These numbers are often referred to as the *indices* of the epikernel or kernel symmetries. Hence the index of D_{4h} in O_h is 3, while the index of D_{2h} is 6. These indices indeed correspond to the numbers of equivalent epikernel or kernel configurations in Fig. 1.

The trial and error procedure, which we have been following to scan the distortion space of a degenerate representation, can be bypassed by a rigorous group theoretical method, based on the genealogy of the parent group. This method has been described in detail elsewhere [3]. Here we will only list a few representative results in Table 2. More extensive epikernel charts may be found in the literature [4, 6, 15, 17]. We also remark that the drawings of normal modes in Herzberg's monograph [18] on IR spectra, often show canonical components, which conserve epikernel groups. Finally several properties of epikernels are collected in the paper by Murray-Rust, Bürgi and Dunitz [6].

Table 2 also shows that there are often several epikernels corresponding to one specific group G and one specific representation Λ. These epikernels can represent independent ways of symmetry lowering, leading to the kernel group along different paths. For example the e″ distortions in D_{3h} can give rise to

$$D_{3h} \begin{array}{c}\nearrow\\ \searrow\end{array} \begin{array}{c}C_2\\ C_s(\sigma_v)\end{array} \begin{array}{c}\searrow\\ \nearrow\end{array} C_1$$

Table 2. Kernel and epikernel subgroups of degenerate representations[a]

Doubly degenerate representations

Class I, n_E = odd					Class II, n_E = even				
G	Λ	K(G,Λ)	E(G,Λ)	n_E	G	Λ	K(G,Λ)	E(G,Λ)	n_E
D_3	e	C_1	C_2	3	D_4	e	C_1	$C_2(C_2'), C_2(C_2'')$	4
D_5	e_1,e_2	C_1	C_2	5	D_6	e_1	C_1	$C_2(C_2'), C_2(C_2'')$	6
D_6	e_2	$C_2(C_6^3)$	D_2	3	C_{4v}	e	C_1	$C_s(\sigma_v), C_s(\sigma_d)$	4
D_7	e_1,e_2,e_3	C_1	C_2	7	C_{6v}	e_1	C_1	$C_s(\sigma_v), C_s(\sigma_d)$	6
C_{3v}	e	C_1	C_s	3	D_{3h}	e″	C_1	$C_2, C_s(\sigma_v)$	6
C_{5v}	e_1,e_2	C_1	C_s	5	D_{4h}	e_g	C_i	$C_{2h}(C_2'), C_{2h}(C_2'')$	4
C_{6v}	e_2	C_2	C_{2v}	3		e_u	$C_s(\sigma_h)$	$C_{2v}(C_2'), C_{2v}(C_2'')$	4
D_{3h}	e′	$C_s(\sigma_h)$	C_{2v}	3	D_{5h}	e_1'', e_2''	C_1	$C_2, C_s(\sigma_v)$	10
D_{5h}	e_1',e_2'	$C_s(\sigma_h)$	C_{2v}	5	D_{6h}	e_{1g}	C_i	$C_{2h}(C_2'), C_{2h}(C_2'')$	6
D_{6h}	e_{2g}	$C_2(C_6^3)$	D_{2h}	3		e_{1u}	$C_s(\sigma_h)$	$C_{2v}(C_2'), C_{2v}(C_2'')$	6
D_{3d}	e_g	C_i	C_{2h}	3		e_{2u}	$C_2(C_6^3)$	$C_{2v}(\sigma_v), D_2$	6
D_{5d}	e_{1g},e_{2g}	C_i	C_{2h}	5	D_{2d}	e	C_1	$C_2(C_2'), C_s(\sigma_d)$	4
D_{6d}	e_4	S_4	D_{2d}	3	D_{3d}	e_u	C_1	C_2, C_s	6
T_d	e	D_2	D_{2d}	3	D_{4d}	e_1, e_3	C_1	$C_2(C_2'), C_s$	8
O	e	$D_2(C_4^2)$	D_4	3		e_2	$C_2(C_4^2)$	C_{2v}, D_2	4
O_h	e_g	$D_{2h}(C_4^2)$	D_{4h}	3	D_{5d}	e_{1u},e_{2u}	C_1	C_2, C_s	10

Triply degenerate representations

G	Λ	K(G,Λ)	E(G,Λ)
T_d	t_1	C_1	C_3, S_4, C_s
	t_2	C_1	C_{3v}, C_{2v}, C_s
O_h	t_{1g}	C_i	$S_6, C_{4h}, C_{2h}(C_2)$
	t_{2g}	C_i	$D_{3d}, D_{2h}(C_4^2, C_2), C_{2h}(C_2)$
	t_{1u}	C_1	$C_{3v}, C_{4v}, C_{2v}(C_2), C_s(\sigma_d), C_s(\sigma_h)$
	t_{2u}	C_1	$D_3, D_{2d}(C_4^2, C_2), C_{2v}(C_2), C_2(C_2), C_s(\sigma_h)$

Icosahedral fourfold degenerate representation

I_h	g_g	C_i	T_h, D_{3d}, S_6, C_{2h}

Icosahedral fivefold degenerate representation

I_h	h_g	C_i	$D_{5d}, D_{3d}, D_{2h}, C_{2h}$

[a] G is the parent group and Λ is the irreducible representation. K(G,Λ) and E(G,Λ) denote the kernel and epikernels; n_E is the index of the epikernels. As can be seen the doubly degenerate representations may be grouped into two classes, depending on even or odd values of n_E. Representations in class I have one unique epikernel of odd index. Representations in class II have two isomorphic epikernels of even index.

Inspection of the epikernels of the *twofold* degenerate representations in Table 2, reveals that these representations can be divided into two classes, depending on the indices of these epikernels. Representations of class I have only one unique type of epikernel, which is always of odd order. Representations of class II have always two different but isomorphic epikernels of even index. As will be shown in Sect. 5, this division into two classes corresponds to two different structural types of JT effect in doubly degenerate electronic states.

Another remark concerns the epikernels of the *threefold* degenerate T representations of T_d and O_h. As may be seen from Table 2, some of these representations give rise to a chain of consecutive subgroups between the parent group and the kernel. In such a chain (see Eq. 11) the epikernel E′ (G, Λ), which is contained in E (G, Λ), is called a *lower ranking*

$$G \supset E(G, \Lambda) \supset E'(G, \Lambda) \supset K(G, \Lambda) \tag{11}$$

epikernel with respect to E (G, Λ). As an example for the tetrahedral t_2 representation, C_s is a lower ranking epikernel with respect to both C_{2v} and C_{3v}.

$$T_d \begin{smallmatrix} \nearrow & C_{3v} & \searrow \\ \searrow & C_{2v} & \nearrow \end{smallmatrix} C_s \rightarrow C_1$$

3.3 The Coordinate Representation is Reducible

It is possible that the coordinates of the nuclear displacement under consideration cover more than one symmetry representation. The corresponding N-dimensional coordinate representation is then said to be reducible. The procedure to find the conserved subgroups in a reducible distortion space closely resembles the treatment for a degenerate irreducible space [3]. Symmetry elements, which leave the entire set of coordinates invariant, constitute the kernel of the space. The kernel group will be conserved in each direction, that is, under an arbitrary displacement, describable as a linear combination of the N coordinates. The elements of the kernel group may immediately be traced by collecting the elements with the same character as the identity operation. In addition the distortion space will contain privileged directions or subspaces, which preserve more symmetry elements and correspond to epikernel groups. As before the indices of these epikernels are given by the quotient of the relevant group orders.

An obvious feature of a reducible distortion space, is that it can in fact be decomposed into irreducible subspaces. The conserved subgroups of these irreducible subspaces may immediately be determined from Tables 1 and 2. Since all these resulting subgroups are conserved in a *part* of the total distortion space, they will all be *epikernels* of the total space. Symmetry reduction thus facilitates the search for epikernels in reducible spaces.

As an example we may study the $b_{1g} + b_{2g}$ distortion space, which forms the JT active space of a twofold degenerate electronic E state in D_{4h} symmetry. The kernel of this combined space is C_{2h}, as may immediately be verified from the following character string

D_{4h}	E	$2C_4$	C_4^2	$2C_2'$	$2C_2''$	i	$2S_4$	σ_h	$2\sigma_v$	$2\sigma_d$
$b_{1g} + b_{2g}$	2	−2	2	0	0	2	−2	2	0	0

On the other hand the kernels of the irreducible subspace b_{1g} and b_{2g} are respectively $D_{2h}(C_2')$ and $D_{2h}(C_2'')$ as indicated in Table 1. The symmetry of the b_{1g} subspace, i.e. $D_{2h}(C_2')$, is not conserved if the b_{2g} distortion is activated; similarly the $D_{2h}(C_2'')$ symmetry is lowered to C_{2h} if the b_{1g} distortion is activated. The subspace kernels therefore become epikernels of the sum representation. We thus have:

$$
\begin{aligned}
C_{2h} &= K(D_{4h}, b_{1g} + b_{2g}) \\
D_{2h}(C_2') &= E(D_{4h}, b_{1g} + b_{2g}) \\
D_{2h}(C_2'') &= E(D_{4h}, b_{1g} + b_{2g})
\end{aligned}
\tag{12}
$$

Figure 2 presents the spatial distribution of these subgroups. Both D_{2h} epikernels lie along perpendicular coordinate axes. Opposite sides of these axes give rise to equivalent distortions, in accordance with the conclusions of the preceding Sect. 3.1. The index of both D_{2h} epikernels is thus equal to two, and obeys the quotient rule in Eq. (9). In between the epikernel directions lies a combination region, comprising all structures, which result from *simultaneous* b_{1g} and b_{2g} distortions. This region only conserves the kernel group C_{2h} and does not contain any additional epikernels.

A more intricate example of a reducible distortion space is the space of the JT active coordinates for a threefold degenerate T state in T_d or O_h symmetry. Following the first order JT selection rule (Eq. 1), this space comprises distortion modes of e and t_2 symmetry:

$$
\begin{aligned}
&T_d: T = T_1, T_2 && [T]^2 - a_1 = e + t_2 \\
&O_h: T = T_{1g}, T_{2g}, T_{1u}, T_{2u} && [T]^2 - a_{1g} = e_g + t_{2g}
\end{aligned}
\tag{13}
$$

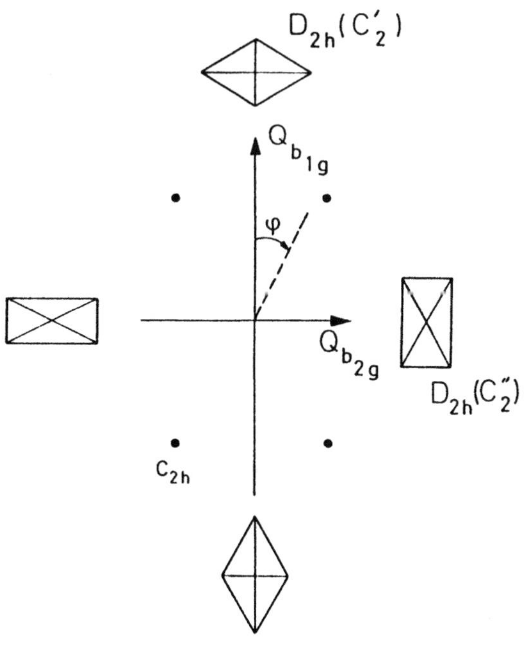

Fig. 2. Schematic representation of the $b_{1g} + b_{2g}$ distortion space of a square planar molecule. The center of the figure is the parent D_{4h} structure. There are two different D_{2h} epikernel symmetries: points along the $Q_{b_{1g}}$ distortion coordinate refer to rhombic structures with $D_{2h}(C_2')$ symmetry, while points on the $Q_{b_{2g}}$ distortion coordinate represent rectangular structures with $D_{2h}(C_2'')$ symmetry. All other points correspond to configurations with C_{2h} kernel symmetry. Remark that the $Q_{b_{1g}}$ and $-Q_{b_{1g}}$ distortions give rise to equivalent structures, and similary for the $Q_{b_{2g}}$ and $-Q_{b_{2g}}$ distortions. Hence each epikernel point has a multiplicity of two; on the other hand, for each kernel point the number of equivalent distorted configurations is four – as illustrated by the *black circles*

Table 3. Kernel and epikernel subgroups for the reducible $e + t_2$ (or $e_g + t_{2g}$) representation in tetrahedral (or octahedral) symmetry[a]

G	Λ[b]	K(G,Λ)	E(G,Λ)
T_d	e	D_2	D_{2d}
	t_2	C_1	C_{3v}, C_{2v}, C_s
	(e,t_2)	C_1	C_{2v}, C_s, C_2
O_h	e_g	$D_{2h}(C_4^2)$	D_{4h}
	t_{2g}	C_i	D_{3d}, $D_{2h}(C_4^2,C_2)$, $C_{2h}(C_2)$
	(e_g,t_{2g})	C_i	$D_{2h}(C_4^2,C_2)$, $C_{2h}(C_2)$, $C_{2h}(C_4^2)$

[a] G is the parent group, Λ is the representation; K(G,Λ) and E(G,Λ) denote the kernel and epikernel subgroups.

[b] The distortion space, with reducible symmetry Λ, is divided into three parts: e,t_2, (e,t_2) for T_d, and similarly for O_h. The e and t_2 part refer to configurations that are obtained by displacements of pure e or t_2 symmetry. The (e,t_2) part contains all configurations that are obtained by combinations of e and t_2 displacements.

The structure of this five-dimensional reducible space has been discussed at length in a previous paper [3]. Table 3 summarizes the results. In the table the total sum $e + t_2$ is divided into three parts: e, t_2 and (e, t_2). The first two parts refer to distorted configurations, which are obtained by displacements of pure e or t_2 symmetry. Evidently the conserved subgroups in these parts of configuration space may immediately be obtained from Table 2. The third part, denoted (e, t_2) or (e_g, t_{2g}), refers to all distorted structures resulting from a *combination* of e and t_2 modes. In contrast to the previous $b_{1g} + b_{2g}$ example, such distortions of mixed e and t_2 symmetry type are seen to give rise to not less than three different epikernels: C_{2v}, C_s and C_2 in T_d, or the holohedrized analogues in O_h. The C_{2v} and C_s epikernels also occur in the irreducible t_2 space, but the C_2 epikernel is truly characteristic for the (e, t_2) distortion of mixed symmetry type. Clearly in the (e, t_2) part of space C_{2v} is a higher ranking epikernel than C_s or C_2.

Finally a special case arises in cyclic groups such as C_n, S_n, C_{nh} and in the tetrahedral groups T and T_h, which contain non-degenerate *complex* representations. Such representations always occur in degenerate pairs with conjugate characters and hence form a reducible space of dimension two. The kernel of this space may again easily be determined from its character set. However, since both irreducible components have complex conjugate transformational properties, it is impossible to find a subgroup, which leaves one component invariant, while transforming the other one. As a consequence the distortion space of complex conjugate representations will be devoid of epikernel directions.

4 The Epikernel Principle

From this section onward we will only be concerned with distortion spaces in the neighborhood of a JT origin. First we will discuss the *epikernel principle* and its general group theoretical basis. The next section deals with the specific implications of the principle for

JT activity in twofold and threefold degenerate states. Subsequently some experimental manifestations of the principle in various areas of chemistry are presented. The final section emphasizes the uniform structure of all JT problems.

Before presenting the epikernel principle, we will briefly recall the essential features of the JT effect. For more details the reader is referred to the monograph of Bersuker [1]. JT instability requires the presence of an orbitally degenerate electronic state. The symmetry representation of this state is labeled Γ, with components $|\Gamma\gamma_i>$, $|\Gamma\gamma_j>$... The JT distortion space consists of symmetry coordinates $Q_{\Lambda\lambda}$, that obey the first order JT selection rule in Eq. (1). In the preceding section we investigated the relationship between the representational symmetry of the distortion coordinates and their symmetry lowering capacity, using the concepts of kernel and epikernel. At present we are interested in the construction of a potential energy surface, which will tell us how the electronic energy of the system changes as a function of the $Q_{\Lambda\lambda}$ distortion coordinates. Since such displacements remove the initial electronic degeneracy, the corresponding potential energy surface cannot be single valued, but will consist of several sheets, which intersect at the point of degeneracy. The number of sheets equals the dimension of Γ. According to perturbation theory [19], these sheets may be obtained by diagonalizing a perturbation matrix $\mathbb{H}$, which is a function of the $Q_{\Lambda\lambda}$ coordinates. One has for the ij element:

$$H_{ij} = \delta_{ij} \sum_{\Lambda\lambda} 1/2\, K_{\Lambda} Q_{\Lambda\lambda}^2 + \sum_{\Lambda\lambda} < \Gamma\gamma_i | \partial\mathcal{H}/\partial Q_{\Lambda\lambda} | \Gamma\gamma_j >_0 Q_{\Lambda\lambda} \tag{14}$$

Here δ_{ij} is the Kronecker delta, $\Lambda\lambda$ runs over the JT active coordinates, and i and j cover the set of $\Lambda\lambda$ components. The parameter K_{Λ} represents the harmonic force constant of the Λ mode. $\mathcal{H}$ is the electronic hamiltonian. The matrix element containing the derivative of $\mathcal{H}$ versus $Q_{\Lambda\lambda}$ must be evaluated at the origin of coordinate space, as indicated by the zero subscript. It must be realized that the expression in Eq. (14) only contains the most essential contributions, viz. the *harmonic* terms and the *linear* terms. Further refinements include other quadratic interactions, such as the *bilinear* terms in $Q_{\Lambda\lambda} \times Q_{\Lambda'\lambda'}$, or even higher order terms.

Diagonalization of the $\mathbb{H}$ matrix as a function of the $Q_{\Lambda\lambda}$'s will yield a multivalued JT potential energy surface. The harmonic restoring forces, described by the quadratic terms in Eq. (14), tend to stabilize the highly symmetrical parent configuration at the center of the surface. In contrast the linear terms represent the distorting forces which cause the instability of the parent configuration with respect to symmetry lowering distortions. The first-order JT selection rule in Eq. 1 derives from the symmetry constraints of these terms [20].

The energy difference between the JT origin and the absolute minimum of the surrounding energy surface is called the *JT stabilization energy*. As could be expected this energy will increase with increasing distorting forces, and decrease with increasing harmonic force constants. Several theoretical methods have been used to calculate the JT stabilization energies for a variety of JT systems. These methods range from simple model treatments to extensive ab initio calculations [2].

These studies have revealed that the JT distortions, which lead to the stable minimum of the potential energy surface, tend to preserve as many symmetry elements as possible. On the basis of such observations, Liehr conjectured that the symmetry of the stable JT geometry would be the highest, which is yet compatible with the loss of initial electronic degeneracy [21]. A more general and more precise description of the symmetry characteristics of JT instabilities may be based on the concepts of kernels and epikernels. Accordingly, in 1984, we formulated the following epikernel principle [3]:

Extremum points on a JT surface prefer epikernels to kernels; they prefer maximal epikernels to lower ranking ones. As a rule stable minima are to be found with structures of maximal epikernel symmetry.

The epikernel principle is more general than Liehr's conjecture in that it also points to the symmetry conservation in non-minimal extrema, such as hill tops and saddle points. Furthermore it is more precise, in that it allows the association of stable minima with high ranking epikernels, which do not obey Liehr's conjecture. A case in point are the orthorhombic minima of the cubic T instability (vide infra, Sect. 5.2). Last but not least, the epikernel principle is not conjectural but derives from the fundamental properties of JT potentials [3, 4]. This important aspect will now be clarified.

Although the JT theorem and the epikernel principle emphasize antithetical aspects, they are not contradictory. Indeed both rules apply to different regions of the JT surface; more specifically the symmetry breaking theorem only concerns configurations in degenerate states, while the epikernel principle discusses the stationarity of configurations in non-degenerate states. Quite remarkably the same symmetry selection rule (Eq. 1), which governs the instability of degenerate states, also puts constraints on the stationarity of non-degenerate states. If we apply Eq. (1) to a non-degenerate state Γ_0, the symmetrized square $[\Gamma_0]^2$ is seen to be totally symmetric so that $[\Gamma_0]^2 - A_1$ is empty. Therefore, *if* the system happens to be in an epikernel configuration – in a non-degenerate state – there can be no symmetry lowering forces whatsover, driving the system out of this epikernel configuration into a surrounding kernel region. In contrast, kernel configurations can be subject to forces that are oriented towards epikernel configurations, since such symmetry increasing forces are totally symmetric under the point group operations of the kernel symmetry and therefore their existence does not violate the selection rule. As a result, non-degenerate epikernel states are indeed expected to be among the stationary points of the surface, and the more so, the higher their ranking [3].

The epikernel principle is a stronger principle though in that it challenges the possibility of kernel minima. A detailed group-theoretical analysis of this second aspect of the principle has been presented in Ref. 4. Here we briefly recall the essential conclusions from this treatment. First of all it must be realized that for each kernel point there will be n_K equivalent kernel points, exactly as each epikernel point has a multiplicity which equals the corresponding epikernel index n_E (see Eqs. 9, 10), However, since $K(G, \Lambda)$ is a subgroup of $E(G, \Lambda)$ (Eq. 7), it follows that n_K is a multiple of n_E. Hence kernel extrema, if they exist, will be more numerous than epikernel extrema of a given type. In order to be stationary at all these equivalent points, the JT surface must be of considerable complexity. As we have shown in Ref. 4, the linear and second order terms in the perturbation matrix of Eq. (14) are unable to impart this degree of complexity to the surface. Only higher order terms in the expansion are able to generate non-symmetrical extrema, provided that the corresponding vibronic coupling constants exceed certain critical values, as compared to the lower order coupling constants. *However – from a perturbational point of view – such a dominance of higher order terms over first and second order contributions is extremely unlikely.* This observation, in conjunction whith the earlier symmetry considerations, rationalizes the epikernel principle.

In conclusion the epikernel principle appears to be related to the specific symmetry lowering properties of the principal terms in the JT hamiltonian. As such it is clear that it only applies to distortions of molecules in JT unstable degenerate states. The next section will present an overview of the typical JT problems, to which the epikernel principle refers.

5 JT Surfaces for Electronic Doublets and Triplets

In this section we will investigate the different adiabatic surfaces for electronic orbital doublets and triplets. The treatment is limited to so-called *ideal* JT systems, i.e. systems with only one JT active mode for each allowed symmetry representation of $[\Gamma]^2 - A_1$. Although non-ideal or multimode JT problems give rise to more complicated surfaces, it is interesting to note that their extremal properties are similar to those of the corresponding ideal problems [22]. Hence as far as the applicability of the epikernel principle is concerned, the neglect of the multimode effect is not expected to affect the conclusions of the present treatment.

5.1 Electronic Doublets

Ideally electronic E states couple to only two JT active coordinates. These may be degenerate as in the case of the cubic $E \times e$ problem (Fig. 1), or non-degenerate, as in the tetragonal $E \times (b_1 + b_2)$ problem (Fig. 2). Both types of problem have been studied in great detail [1, 2]. Here we review these solutions, with special emphasis on their relationship with the epikernel principle.

5.1.1 The Cubic $E \times e$ Problem

If the JT perturbation matrix for the cubic $E \times e$ problem is limited to the linear terms in Q_ε and Q_θ and the harmonic term in $Q_\theta^2 + Q_\varepsilon^2$, the resulting adiabatic surface adopts the well known rotational shape of a mexican hat (see Fig. 3). On this surface one immediately recognizes a circular trough surrounding the conical intersection at the JT origin. This highly symmetrical result is of course a limiting case. A more realistic potential must include the remaining quadratic or bilinear terms in $Q_\theta^2 - Q_\varepsilon^2$ and $Q_\theta Q_\varepsilon$. Upon introduction of these terms the circular trough will become distorted or warped. The resulting potential still contains a continuous minimal energy valley, though no longer of uniform depth. The deformation of the trough potential as a function of the angular coordinate of the distortion space may be described by a series of *warping terms*. These consist of an angular function times a weight factor, the so-called *vibronic coupling constant*. When adding the warping terms to the trough potential, one must keep in mind the trigonal symmetry of the underlying configuration space shown in Fig. 1. Epikernel configurations cannot be unstable with respect to neighbouring kernel configurations. Likewise equivalent configurations cannot have different potential energies. Only angular functions characterized by the trigonal symmetry of the configuration space (Fig. 1) – i.e. so-called trigonal *invariants* – can meet these requirements.

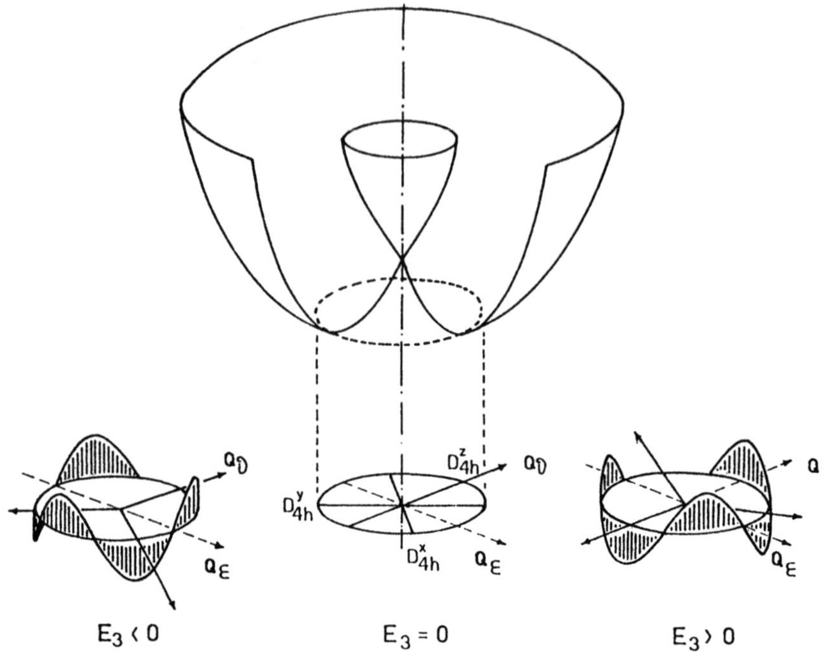

Fig. 3. The circular mexican hat surface of the cubic E × e problem. The minimal energy trough of this surface forms a circular valley around the intersection point. The lower half of the figure schematically represents the warping of this trough as a result of non-harmonic interactions. E_3 is the associated vibronic coupling constant (see Eq. 15). For $E_3 = 0$ there is no warping effect and the trough simply corresponds to an equipotential circle in (Q_θ, Q_ε) space. For $E_3 < 0$, minima will be found along the positive Q_θ axis and at angles of 120° and 240°. For $E_3 > 0$, minima will be found in opposite directions, i.e. along the negative Q_θ axis and at angles of 60° and 300°

As we have demonstrated elsewhere [4], the *leading* warping term of the cubic E × e problem is given by the following expression:

$$\mathcal{W}_3 = E_3 \frac{1}{\sqrt{\pi}} \cos 3\varphi \tag{15}$$

Here E_3 is the vibronic coupling constant and $\frac{1}{\sqrt{\pi}} \cos 3\varphi$ is a normalized cylindrical harmonic of the third rank. The φ coordinate is defined in Fig. 1. $\mathcal{W}_3$ is the lowest order warping term, which is totally symmetric under the symmetry operations of the e space. Its periodicity of 120° exactly matches the trigonal distribution of the epikernel directions. Its extremal points will determine the stationary configurations of the E × e surface. As shown in Fig. 3 two different solutions are possible, depending on the sign of the coupling constant. One solution ($E_3 < 0$) is to associate the three minima with the positive (or elongated) epikernel configurations. The three maxima then coincide with the negative (or compressed) epikernel configurations in the trough. These therefore become saddle points in a circular path between the elongated epikernel isomers. The other solution is obtained by changing the sign of the warping term. In this case ($E_3 > 0$)

the elongated epikernel points become saddles and the compressed ones become minima. For either possibility no extra folds are left in between the epikernels, which means that there are no stationary kernel configurations. The first warping term which would stabilize kernel configurations, is a component transforming as cos 6φ. In chemical applications the contribution of this term is negligible in comparison with the cos 3φ component. Exceptions to the epikernel principle are thus unlikely to occur.

5.1.2 The Tetragonal E × (b_1 + b_2) Problem

In the limit of *equal* JT stabilization energies for the b_1 and b_2 modes, the adiabatic potential energy surface of the E × (b_1 + b_2) problem very much resembles the mexican hat profile, as illustrated in Fig. 4.

The only – unimportant – difference is that the equipotential trough, surrounding the intersection, in general will be elliptical, instead of circular. This highly symmetrical solution only exists in the limiting case of accidental degeneracy. Since there is no symmetry basis for equal b_1 and b_2 JT stabilization energies, in general a breakdown of this degeneracy, a so-called *mode splitting,* will be appropriate. As before this introduction of mode splitting may be described as a warping process of the equipotential trough. The leading term of this process transforms as the harmonic component, given in Eq. (16).

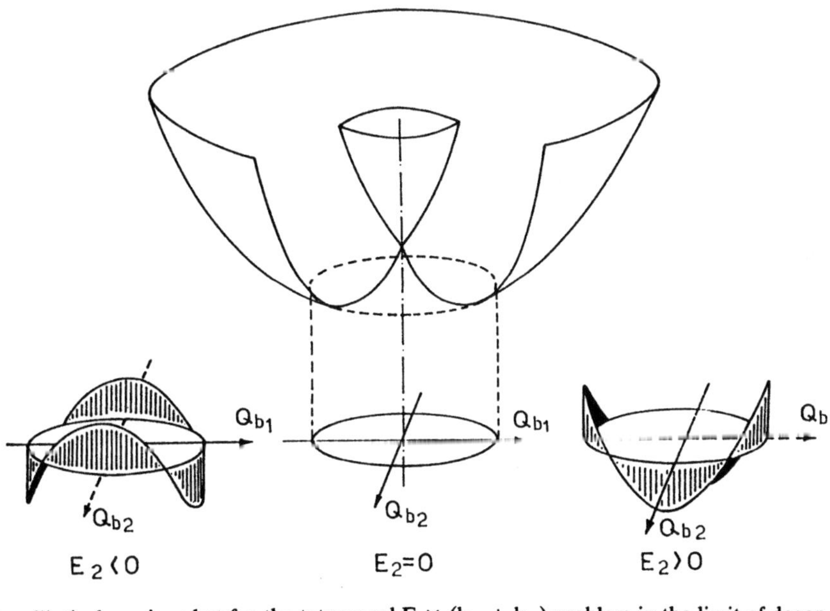

Fig. 4. The elliptical mexican hat for the tetragonal E × (b_{1g} + b_{2g}) problem in the limit of degenerate coupling. The corresponding coordinate space is defined in Fig. 2. The lower half of the figure represents the warping effect of mode splitting. E_2 is the associated vibronic coupling constant (see Eq. 16). For $E_2 = 0$ there is no warping term and the trough simply corresponds to an equipotential ellips, with its principal axes along the $Q_{b_{1g}}$ and $Q_{b_{2g}}$ directions of the coordinate space. For $E_2 < 0$, minima will be found in the rhombic epikernels on the $Q_{b_{1g}}$ coordinate. For $E_2 > 0$ minima will be found in the rectangular epikernel directions along the $Q_{b_{2g}}$ mode

$$\mathcal{W}_2 = E_2 \frac{1}{\sqrt{\pi}} \cos 2\varphi \tag{16}$$

Again this is the lowest rank harmonic, which is compatible with the overall twofold symmetry of the distortion space of Fig. 2. Depending on the sign of the relevant vibronic coupling constant, this term will produce minima in the $D_{2h}(C_2')$ epikernels and saddle points in the alternating $D_{2h}(C_2'')$ epikernels, or vice versa. Due to the minimal rank of the harmonic component, no extremal kernel configurations will be found. Stabilization of such configurations by higher order coupling effects is unlikely, since the bilinear term in $Q_{b_1} \times Q_{b_2}$ cannot contribute to the perturbation hamiltionian for symmetry reasons. These results are in agreement with the epikernel principle.

5.1.3 Generalization

The two examples above are paradigmatic for the two possible structural patterns of JT surfaces for orbital doublets. A useful representation of these examples, which easily lends itself to generalization, consists in drawing a regular polygon, that matches the epikernel structure of the distortion space. The cubic E × e problem, which has a trigonal structure (Fig. 1), is thus represented by a regular triangle, in such a way that the three principal axes of inertia of the triangle coincide with the epikernel directions.

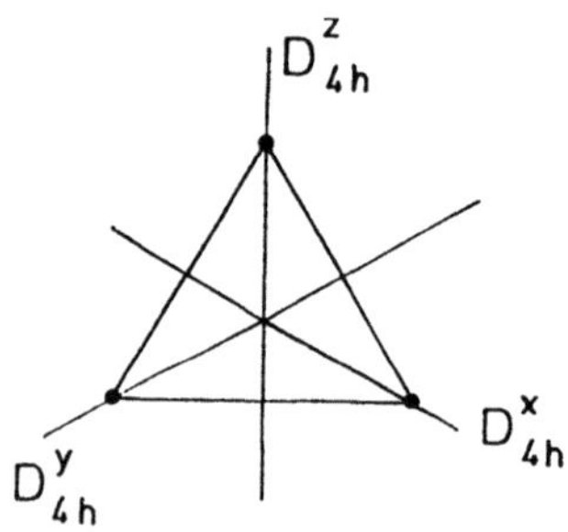

This drawing clarifies two important aspects of the E × e problem. First it shows that there is only one type of epikernel, since the polygon has only one type of principal axis of inertia; second it explains that the epikernel axes lack a centre of inversion, since they link a vertex to an edge. Hence displacements in opposite directions on such an axis will not lead to equivalent structures. In fact on the warped surface, one direction will point to a saddle point, while the opposite direction points to a minimum. These aspects are common to all E type instabilities, which may be represented by a n-polygon for which the number of vertices, n, is odd. In all these cases, there will be only one type of epikernel, with odd index $n_E = n$. The dominant warping term will always be of the lowest allowed rank, namely n, so as to produce minima and saddle points on opposite sides of the epikernel directions. This is the case for all instabilities with active modes transforming as class I representations in Table 2. These coordinates do indeed have one unique epikernel of odd index.

The other category of E type problems comprises all instabilities which must be represented by *even* n-polygons. Such polygons differ from *odd* n-polygons in that they have two separate sets of principal axes, one between opposite vertices and one between opposite edges. Furthermore each of these axes has inversion symmetry. Hence in the corresponding distortion space there will be two different types of epikernels, both with even indices. Opposite epikernel directions will give rise to equivalent distortions. The dominant warping term of rank n will associate minima with one type of epikernels and saddle points with the other type. An example with n = 4 is the $E_1 \times e_2$ problem in D_{4d} symmetry. The e_2 distortion space can be represented by a square. The two $E(D_{4d}, e_2)$ epikernels are C_{2v} and D_2 (Cf. Table 2). The corresponding distortion coordinates may be oriented along the two sets of principal axes of the square, as shown below.

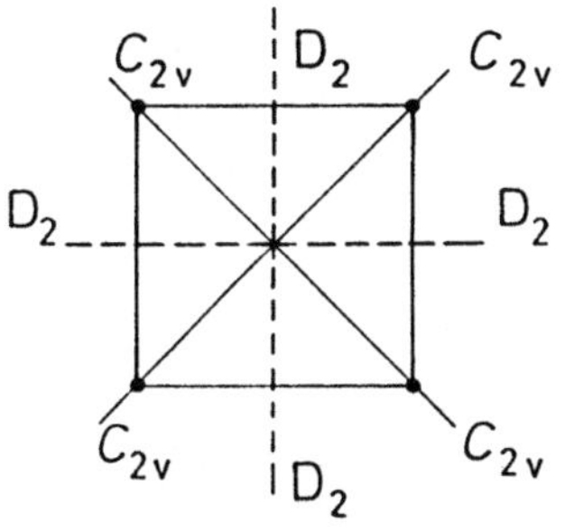

The even n catergory further applies to all coordinates, which transform as class II representations in Table 2. In fact the reducible distortion space of $b_1 + b_2$ symmetry may formally by characterized as a class II representation too, having two different but isomorphic epikernels of even index. The results of the tetragonal $E \times (b_1 + b_2)$ problem may thus be represented by the trivial 2-polygon. This polygon consists of a single line joining two vertices. It had indeed two principal axes, as shown below. These axes symbolize the two different D_{2h} epikernel directions of the $b_1 + b_2$ space (Cf. Fig. 2).

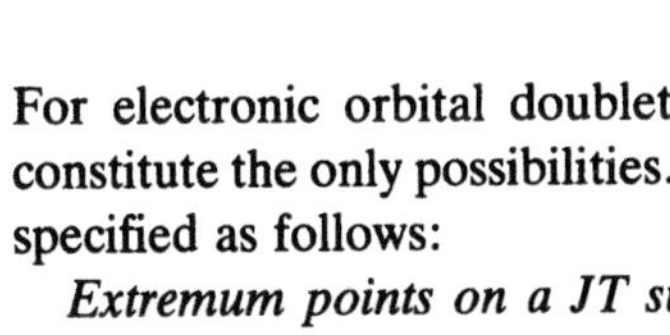

For electronic orbital doublets the two patterns, corresponding to odd and even n, constitute the only possibilities. Hence the epikernel principle for E terms may be further specified as follows:

Extremum points on a JT surface for an orbital doublet will coincide with epikernel configurations. If the distortion space conserves only one type of epikernel, minima and saddle points will be found on opposite sides of the same epikernel distortion. If the distortion space conserves two types of epikernels, minima and saddle points will be characterized by different epikernel symmetries.

5.2 Electronic Triplets

Electronic orbital triplets or T terms only occur in the cubic and icosahedral groups. Ideally these terms couple to five JT active coordinates. For the icosahedral T states, the five JT coordinates transform according to one single fivefold degenerate representation, which is labeled h_g (in I_h) or h (in I) (see Eq. 17).

$$\begin{array}{lll} I\ : T = T_1, T_2 & [T]^2 - a\ = h & \\ I_h : T = T_{1g}, T_{2g}, T_{1u}, T_{2u} & [T]^2 - a_g = h_g & \end{array} \qquad (17)$$

In the cubic case the symmetry of the JT modes follows the reducible representation $e + t_2$ (in T_d and O) or $e_g + t_{2g}$ (in O_h) (see Eq. 13). The epikernels of these JT coordinates may be found in Tables 2 and 3. In view of the high dimensionality of the distortion spaces it is difficult to visualize the spatial distribution of these epikernels, not to mention the pictorial representation of a complicated potential surface in the corresponding distortion space. One possibility is to draw cross-sections of the most relevant subspaces [3]. Another more comprehensive alternative, wich combines all essential features in one drawing, may be realized by using a projection technique. The projection space is the space of the electronic functions. This space may be defined by three cartesian axes which represent the three components $|Tx>$, $|Ty>$ and $|Tz>$ of the electronic T state. An eigenfunction, say $|Ta>$, of the distorted hamiltonian $\mathbb{H}$ (Eq. 14) can be expressed as a linear combination of these three basis functions.

$$|Ta> = a_x|Tx> + a_y|Ty> + a_z|Tz> \qquad (18)$$

with: $a_x^2 + a_y^2 + a_z = 1$

In the three-dimensional function space, $|Ta>$ can be associated with a coordinate point (a_x, a_y, a_z). Since these coefficients are normalized, all such points will lie on a unit sphere. Angular displacements on this sphere may be conceived as changes in the composition of the wave function. In order to represent the JT suface, we will focus our attention on the lowest eigenvalue of $\mathbb{H}$. This eigenvalue corresponds to a sheet in the five-dimensional coordinate space. To each point in this space corresponds a well-defined distortion, and at the same time a well specified ground state energy, and ground state wavefunction. Obviously the mapping between these points in distortion space and the associated wavefunctions in function space must be of a many-to-one type, in view of the lower dimensionality of the function space. Since only one energy value can be represented for each function point, a selection criterion is needed. The most relevant drawing is obtained by applying the following mapping principle:

In each point (a_x, a_y, a_z) of the globe, we plot the lowest eigenvalue, corresponding to the electronic eigenfunction $|Ta>$ [23].

In order to avoid the difficulties of visualizing the energy variations along an axis, perpendicular to the globe, one can display iso-energy contours on the spherical surface. In this way one obtains a projection in three dimensions of the low energy regions of the full five-dimensional JT surface. Such a plot will show in a concise way the spatial distribution of absolute minima and least energy paths. From a chemical point of view, these are the two most important structural features of the adiabatic surface. This plotting technique will now be used to represent the warping effects for icosahedral and octahedral T terms.

5.2.1 *The Icosahedral T × h Problem*

The icosahedral T × h problem has been studied by Khlopin, Polinger and Bersuker [24], considering only linear and harmonic terms. In this approximation, the projection of the JT surface onto function space simply yields a constant spherical potential. This means that the minimal energy region of the JT surface consists of a continuous spherical equipotential trough, which surrounds the JT origin. The situation is entirely analogous to the circular equipotential trough, which encapsulates the conical intersection in the cubic E × e problem (Cf. Fig. 3). A more elaborate JT potential, which includes the remaining quadratic terms, will give rise to distortion and warping of the equipotential trough. On the unit sphere this warping effect may directly be described by suitable spherical harmonic functions. The leading term of this series will be the lowest rank spherical harmonic, which is totally symmetric under the icosahedral symmetry of the T × h problem. The resulting warping term [25] of rank 6 is given in Eq. (19), where, x, y and z stand for a_x, a_y and a_z:

$$\mathcal{W}_6 = E_6 \frac{21\sqrt{143}}{80\sqrt{\pi}}\left[z^6 + x^6 + y^6 - \frac{17}{8}(z^4 + x^4 + y^4) - \frac{67}{4}x^2y^2z^2 + \frac{169}{168} - \frac{5\sqrt{5}}{4}(z^2 - x^2)(x^2 - y^2)(y^2 - z^2)\right] \tag{19}$$

with : $x^2 + y^2 + z^2 = 1$

A contour set of this expression is represented [26] in Fig. 5. As the figure shows there are two ways in which this term can be laid over the distortion space. One way ($E_6 < 0$) is by associating the minima of this function with the pentagonal epikernels (D_{5d}) of the icosahedron. The maxima then coincide with the trigonal epikernels (D_{3d}). As the figure indicates, saddle points will be found at the D_{2h} epikernel configurations. By changing the sign of the vibronic coupling constant E_6, minima and maxima will be interchanged, D_{3d} epikernels now being lower in energy. In any case the epikernel principle is clearly confirmed, in that the highest ranking epikernel configurations will attract the JT instability.

5.2.2 *The Cubic T × (e + t_2) Problem*

Several studies have been devoted to the adiabatic JT hypersurface of the cubic T × (e + t_2) problem [2]. The existence of a spherical equipotential energy trough in the limit of equal JT stabilization energies for the e and t_2 modes, has first been established by O'Brien [27]. As for the icosahedral T term a projective drawing of this trough in function space simply consists of a constant spherical potential.

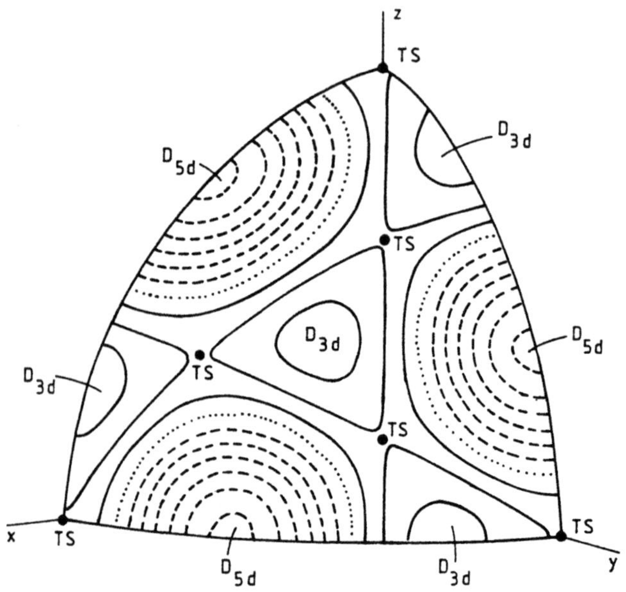

$\mathcal{W}_6\ (E_6 < 0)$

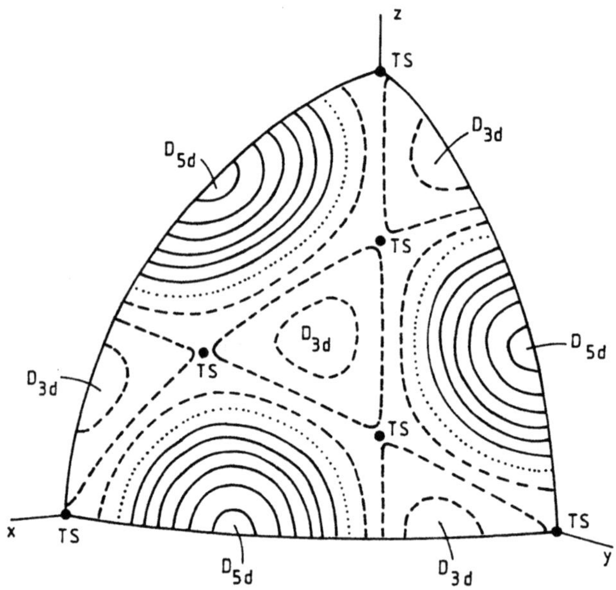

$\mathcal{W}_6\ (E_6 > 0)$

Fig. 5. The icosahedral T × h problem: contour plot of the icosahedral warping term $\mathcal{W}_6$ (see Eq. 19) on the unit sphere in function space. The coordinate frame refers to the three cartesian components of the T state. E_6 is the vibronic coupling constant. Positive, zero, and negative values of $\mathcal{W}_6$ are represented respectively by *full lines, dots,* and *dashed lines.* For $E_6 < 0$, the $\mathcal{W}_6$ function has minima in pentagonal epikernels (D_{5d}) and maxima in trigonal epikernels (D_{3d}). For $E_6 > 0$, minima and maxima are interchanged. In either case the transition states, TS, coincide with D_{2h} epikernel points

Two different symmetry lowering mechanisms may be responsible for the breakdown of this spherical symmetry. One is the splitting of the JT stabilization energies for the e and t_2 modes. The other is the warping effect, due to additional quadratic interaction terms. Both mechanisms have already been encountered in the previous examples, but never simultaneously. Indeed in the tetragonal $E \times (b_1 + b_2)$ problem the surface warping was due to the splitting of the two modes, while quadratic interactions were symmetry forbidden. In contrast, in the single mode problems, $E \times e$ and $T \times h$, mode splitting was excluded but quadratic interactions were symmetry allowed. The cubic T problem is unique in that both effects are simultaneously allowed. Only rather recently was it realized that realistic JT potentials should incorporate both effects [28, 29]. The use of such potentials results in a more flexible warping effect. Indeed, as opposed to the previous examples, the warping of the $T \times (e + t_2)$ trough potential will be controlled by *two* warping terms, instead of one. These terms are components of the $l = 4$ and $l = 6$ spherical harmonics, that are totally symmetric under the octahedral symmetry operations of the problem. The mode splitting alone only requires the $l = 4$ term, but the quadratic interactions also contain the $l = 6$ term. As a result the potential in function space may be approximated as:

$$\mathcal{W} = \mathcal{W}_4 + \mathcal{W}_6 \tag{20}$$

with $\mathcal{W}_4 = E_4 \cdot [(7/12)^{1/2} Y_0^4 + (5/24)^{1/2} (Y_{-4}^4 + Y_4^4)]$

$$\mathcal{W}_6 = E_6 \cdot [-(1/8)^{1/2} Y_0^6 + (7/16)^{1/2} (Y_{-4}^6 + Y_4^6)]$$

$\mathcal{W}_4$ and $\mathcal{W}_6$ are the warping terms of the fourth and sixth tensorial rank. E_4 and E_6 are the associated vibronic coupling constants. (Notice that the crystal field potential for an f-electron in an octahedral lattice is formally equivalent to Eq. 20).

The angular forms of the $l = 4$ and $l = 6$ invariants are depicted [26] in Fig. 6. For $E_4 > 0$, the $l = 4$ invariant has minima and maxima in trigonal and tetragonal points respectively. These assignments can be reversed by changing the sign of the coupling constant E_4. In either case the orthorhombic extrema perform the role of saddle points on the least energy isomerization paths between the stable minima. In contrast the $l = 6$ invariant, which stems from quadratic interactions, has a more pronounced warping effect, in agreement with its higher tensorial rank. Depending on the sign of the accompanying vibronic constant, E_6, it produces minima in the orthorhombic points and hill tops in the tetragonal and trigonal points, or vice versa. The interstitial saddle points are identified as $C_{2h}(C_2)$ structures.

A complete decription of the extremal points for the cubic T instability requires the summation of both warping effects. This summation may give rise to different results, depending on the sign of the vibronic constants E_4 and E_6. Figure 7 indicates the symmetry of the stable minima, as a function of the parameters E_4 and E_6. As can be seen from the figure only the highest ranking epikernels, D_{4h}, D_{3d} and D_{2h}, are eligible. In the border region of two symmetry domains, minima of the adjacent symmetry types may coexist. A special solution is found for parameter values on the borderline between D_{4h} and D_{2h} minima, which is defined by $\sqrt{26}\, E_6 = 20\, E_4/\sqrt{21}$, for $E_4 < 0$. Along this line the minima form a one-dimensional continuum, connecting all tetragonal and orthorhombic points of the phase globe.

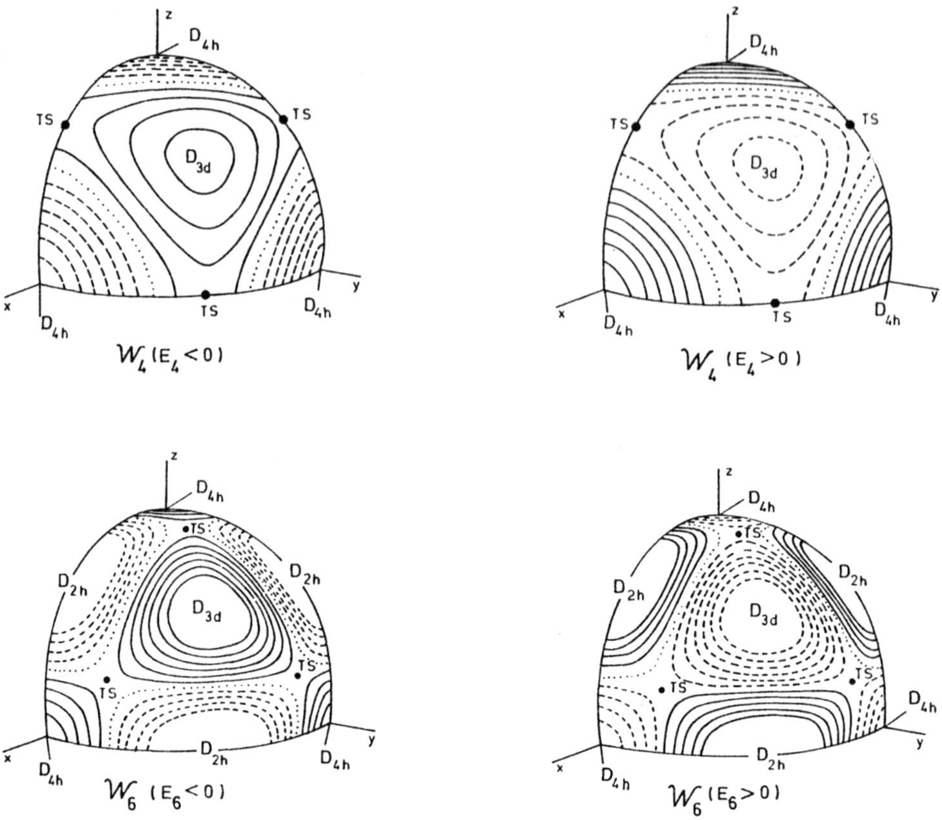

Fig. 6. The cubic T × (e + t_2) problem: contour plots of the octahedral warping terms $\mathcal{W}_4$ and $\mathcal{W}_6$ (see Eq. 20). E_4 and E_6 are the relevant vibronic coupling constants. Positive, zero, and negative values are represented respectively by full lines, dots, and dashed lines. The upper part of the figure represents the fourth rank warping term $\mathcal{W}_4$. This term yields tetragonal ($E_4 < 0$) or trigonal ($E_4 > 0$) minima. In either case the transition states, TS, coincide with the D_{2h} (C_4^2, C_2) epikernels (Cf. Table 3). The lower part displays the sixth rank warping term $\mathcal{W}_6$. This term clearly has a more complex structure. For $E_6 < 0$, it yields minima at the D_{2h} epikernel points; for $E_6 > 0$, both D_{3d} and D_{4h} are stabilized, D_{2h} now being on a hill top. In either case the transition states, TS, coincide with the lower ranking $C_{2h}(C_2)$ epikernels. (In tetrahedral symmetry the D_{4h}, D_{3d}, D_{2h} and $C_{2h}(C_2)$ labels must be replaced respectively by D_{2d}, C_{3v}, C_{2v} and C_s labels)

The solutions in Fig. 7 clarify the application of the epikernel principle to the T × (e + t_2) problem. If the parameter E_6 is small as compared to E_4, the principal symmetry lowering interaction is due to the $l = 4$ warping term. This term represents the splitting of the e and t_2 JT stabilization energies. If the e mode is more stabilized than the t_2 mode ($E_4 < 0$), the stable minima will have tetragonal symmetry, which is the highest ranking epikernel of the e coordinates (Cf. Table 2). If on the other hand the t_2 mode is more stabilized ($E_4 > 0$), the stable minima will have trigonal symmetry, which is the highest ranking epikernel of the t_2 mode (see Table 2). The D_{2h} epikernel is only stabilized for large negative values of E_6. The most favourable coupling mechanism in

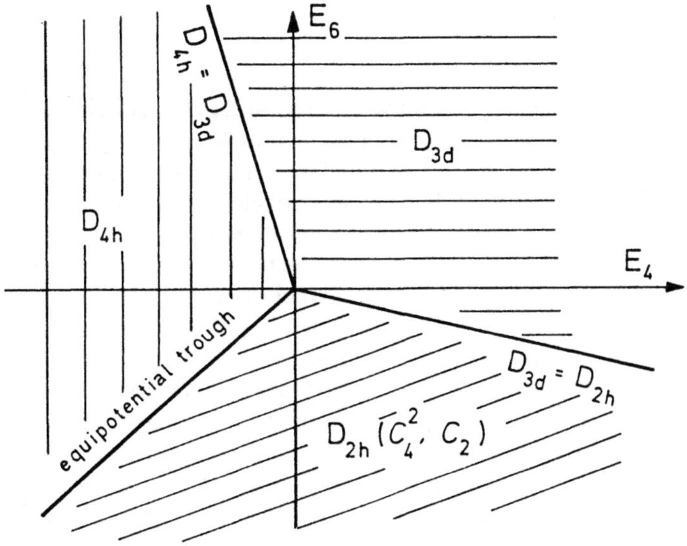

Fig. 7. Cubic T × (e × t_2) problem: domains of existence of absolute minima of a given symmetry type, as a function of the vibronic coupling constants E_4 and E_6. In the border region between two domains, minima of the adjacent symmetry types may coexist. A special solution is found for parameter values on the borderline between D_{4h} and D_{2h} minima: along this line the minima form a continuous mesh, connecting all tetragonal and orthorhombic points

this domain is the bilinear e × t_2 interaction. This interaction reaches its greatest amplitude in the orthorhombic points, which are precisely the highest ranking epikernel sites of the *combined* (e, t_2) space (see Table 3).

Calculations by Bacci [29] et al. and by Bersuker et al. [22, 28] on more elaborate JT potentials confirm these conclusions. Especially noteworthy are the results of Bersuker and Polinger [28], which indicate that the lower ranking $C_{2h}(C_2)$ and $C_{2h}(C_4^2)$ epikernels will never become absolute minima. Furthermore kernel extrema were found to lie beyond the limits of the region of stability of the system. In Ref. 3 the geometry of the JT surface was analyzed, using the classical Oepik and Pryce procedure [19], combined with an extensive examination of the possible pathways between the extremal points on the surface. The conclusions of this paper are fully confirmed by the present projection technique. This technique offers the great advantage that the essential characteristics of the surface, including all three maximal epikernels, can be represented in one single drawing.

6 Applications

In this section we review a few selected examples from the literature, which clearly illustrate the experimental relevance of the epikernel principle.

6.1 Organic Radicals

Conjugated hydrocarbons, in particular cyclic polyenes, offer some of the best documented examples of the JT effect in organic radicals. Kataoka and Nakajima studied the potential energy profiles of such radicals in doubly degenerate E states coupling to e type vibrational modes [30, 31]. These authors concluded that in some cases perturbation theory up to the third power was unable to differentiate between the potential energy profiles along kernel and epikernel directions; in all other cases the potential energy minimum was always found along the mode which preserved the higher symmetry. Never were energy minima found in kernel directions. The conclusions agreed well with the results of extensive ab initio calculations.

The analysis of warping terms provides a simple rationale for these findings. The cases where one can differentiate between kernel and epikernel directions, all have distortion spaces with a trigonal structure. Examples with degenerate ground states are the cyclopropenylradical (C_3H_3), the cation and anion radicals of benzene ($C_6H_6^{\pm}$), and the bicyclo [1,1,1] pentane-2,4,5-triene (C_5H_5). In all these cases a warping term of rank 3 is required to break the rotational symmetry of the degenerate coupling limit (Eq. 15).

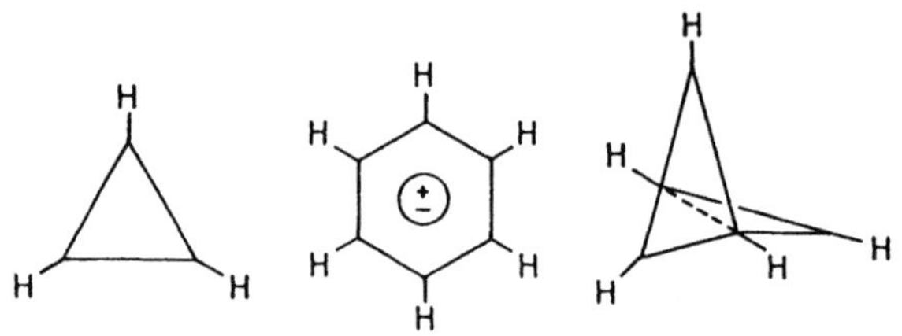

C_3H_3 $C_6H_6^{\pm}$ C_5H_5

Second-order interactions are sufficient to provide such warping terms, exactly as in the case of the cubic E × e problem. On the other hand those examples, where identical energy profiles are obtained along kernel and epikernel directions, involve distortion spaces with a pentagonal, or heptagonal, structure. A case in point is the groundstate of the cyclopentadienyl radical (C_5H_5). The lowest allowed warping terms for these spaces are of rank 5, or 7.

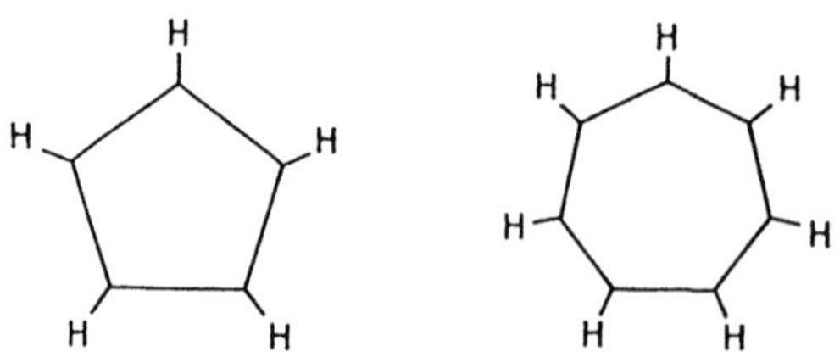

C_5H_5 C_7H_7

In all probability the amplitude of warping terms of this type will be very small, since only higher order coupling mechanisms are able to provide warping terms of such tensorial ranks. As a consequence one indeed expects only minute folds in the energy trough of pentagonal or heptagonal annulene systems. Under these conditions dynamic JT motions will be favored. Similar observations may be made in various metallocenes, consisting of cyclopentadienyl rings coordinated at transition metals [32, 33].

For an example of the tetragonal $E \times (b_1 + b_2)$ type problem we refer to the detailed ab initio calculations on the cyclobutadiene radical cation ($C_4H_4^{+}$) by Borden, Davidson and Feller [34]. The potential energy surface of this cation is found to ressemble the surface in Fig. 4 with rectangular minima and rhomboidal saddle points (i.e. $E_2 > 0$).

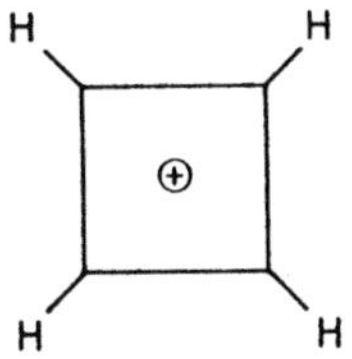

$C_4H_4^{\bullet}$

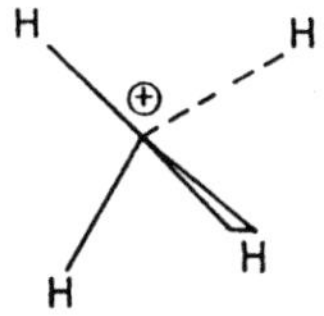

$CH_4^{\bullet}$

The orbital triplet problem $T \times (e + t_2)$ is exemplified by the methane cation radical (CH_4^{+}). Neon matrix ESR measurements have lead to a C_{2v} ground-state assignment, consistent with ab initio calculations [35]. As indicated in Table 3, C_{2v} is the maximal epikernel of all configurations that are obtained by combinations of e and t_2 displacements. In a subsequent study, Pople et al. [36] have explored the full potential energy surface of CH_4^{+} and established the existence of C_s transition states connecting the C_{2v} minima. The representation of this surface in function space may be verified to correspond to the $\mathcal{W}_6$ warping term with $E_6 < 0$, shown in Fig. 6. The C_{2v} symmetry of the absolute minimum was confirmed by Takeshita in a theoretical analysis of the photoelectron spectrum of methane [37]. In addition a local minimum, possibly coexisting, was found in C_{3v} symmetry. Similar distortions appear to be present in the analogous silane, germane and stannane cations [38]. As an example for SiH_4^{+} the C_{2v} structure is calculated to be more stable than the C_{3v} structure [38]. From ESR studies the methylated silane cation, $Si(CH_3)_4^{+}$, indeed appears to have a C_{2v} structure [39].

6.2 Organic Diradicals

Some organic diradicals, for instance the trigonal cyclopropenyl dianion ($C_3H_3^{-}$) or the isoelectronic triaziridenyldication ($N_3H_3^{2+}$), have a low lying $^1E'$ excited state, which exhibits an anomalous type of JT effect [40, 41]. According to the selection rule of

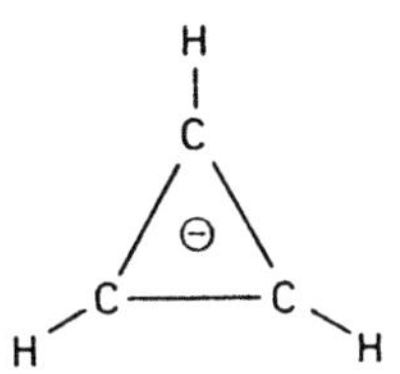

$C_3H_3^{-}$

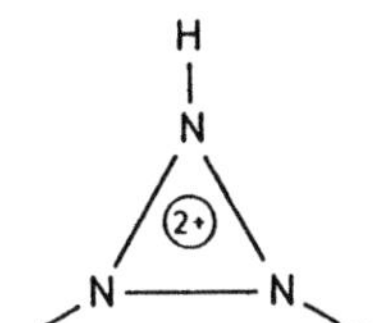

$N_3H_3^{2+}$

Eq. (1), such a trigonal $^1E'$ state couples primarily to an in-plane e′ mode. From the epikernel principle the resulting $E' \times e'$ problem in D_{3h} symmetry is expected to give rise to a C_{2v} distortion, as indicated in Table 2. Nonetheless detailed calculations on $N_3H_3^{2+}$ indicate that the most stable configuration has a non-planar geometry with $C_s\,(\sigma_v)$ symmetry [41]. A similar behaviour is conjectured [40] for $C_3H_3^-$. This C_s symmetry appears to be an epikernel of the out-of-plane e″ mode, as many be seen from Table 2. This vibration is found to be activated by a second order or pseudo-JT effect. In general such a displacement must span a representation Λ belonging to the direct product of the state representation Γ with Γ_i, where Γ_i is the irreducible representation of some suitable low-lying excited state [9, 10].

$$\Lambda \in \Gamma \times \Gamma_i \tag{21}$$

Hence the diradical states are exceptional in that the epikernel principle cannot be based on the usual first order JT section rule of Eq. (1)! The reason is that the $^1E'$ state is essentially based on a half-filled shell configuration, with two electrons in a doubly degenerate e″ orbital. As we have shown elsewhere such half-filled shell states are subject to a *hole electron exchange symmetry, which counteracts the first order JT distorting forces* [42]. Moreover, in these highly reactive molecules the absence of a strong first order activity coincides with a pronounced second order effect, due to suitable low lying excited states. This limits the applicability of the epikernel principle in the case of diradicals.

6.3 Transition-Metal Complexes

Complexes of transition-metal ions by and large represent the most prominent examples of the JT effect. Of special interest are the hexacoordinated complexes of the d^9 Cu^{2+} ion, exemplifying the octahedral $E_g \times e_g$ problem [2]. As indicated in Fig. 1, the e distortion space has an orthorhombic kernel and a tetragonal epikernel. In all cases in which Cu^{2+} is isomorphously substituted into a host lattice site of regular octahedral symmetry, the Jahn-Teller effect induces a tetragonal distortion [43]. Orthorhombic kernel structures are marginal, if at all existent [9]. These observations clearly point to the presence of a trigonal warping term with an appreciable amplitude (Cf. Eq. 15). According to Reinen, Hitchman, and coworkers [44, 45], deviations from this pattern, which may sometimes be observed in less symmetric host lattices, must be attributed to *external* strain effects from the lattice. Such effects can be taken into account by adding an anisotropic strain energy term to the warping potential. It should be remarked that a somewhat different interpretation of the symmetry reducing role of crystal packing effects has been proposed by Hathaway [46]. According to this interpretation, lattice strain does not even perturb the basic structure of the warped mexican hat profile, shown in Fig. 4, but only influences the interconversion rates between the three minima along the low energy valley. In any case both authors agree on the intrinsic trigonal warping of the potential energy surface for isolated CuL_6 complexes, in accordance with the epikernel principle. In this respect one might wonder whether the small orthorhombic distortions of the tetragonal minima that are sometimes observed in CuL_6 crystal structures, can possibly be due to a pseudo-JT distortion of the D_{4h} structure. As far as the pseudo-JT coupling between the two tetragonal E components is concerned, the answer to this question must be negative. Indeed the coupling element, which is responsible for this type of effect, is the linear off-diagonal coupling term between the tetragonal components in the E × e hamiltonian. Clearly this term does not give extra stability to kernel extrema, as the shape of the resulting mexican hat potential illustrates. Hence in all probability the minor deviations from D_{4h} symmetry, that may sometimes occur, should be due to crystal packing forces.

JT instabilities of the T type may best be studied in tetrahedral complexes of d^8 or d^9 metal ions. So far stable minima of these complexes have only been found for D_{2d}, C_{3v} or C_{2v} epikernels in agreement with the analysis in Fig. 7. Fourcoordinated Ni^{2+} and Cu^{2+} complexes usually adopt the D_{2d} structure [47, 48]. In contrast $Fe(CO)_4$ presents a clear example of a C_{2v} distortion [3]. A trigonal geometry must probably be assigned to the matrix-isolated $Co(CO)_4$ fragment [3, 49]. All these symmetry preferences can easily be explained, using the recently developed dynamic angular overlap model [3, 47, 50–52]. CNDO type calculations have also been performed [53, 54].

Interestingly in some of these T type systems, there is evidence for a plasticity effect [48, 55]. Furthermore in the case of $Fe(CO)_4$, interconversion of isomers may be induced by infrared laser irradiation [56]. Such observations corroborate the warped trough model of JT surfaces, on which the epikernel principle is based.

6.4 Cluster Compounds

Boron hydride clusters and their transition-metal analogues comprise a novel class of compounds, which are of great potential importance for the study of the JT effect. In their usual oxidation states, such clusters frequently adopt highly regular polyhedral geometries, obeying simple electron counting rules [57, 58]. Reduction or oxidation of these species may lead to sizeable symmetry lowering distortions.

As an example tetranuclear metal clusters, such as $Ir_4(CO)_{12}$, containing 60 valence electrons, are known to form stable diamagnetic compounds with perfect tetrahedral symmetry. In contrast 62-electron clusters usually exhibit the less symmetric C_{2v} *open butterfly* shape, with one elongated edge [59, 60]. Adding one electron pair to a completely bonding tetrahedron thus gives rise to the cleaving of one edge, as presented below.

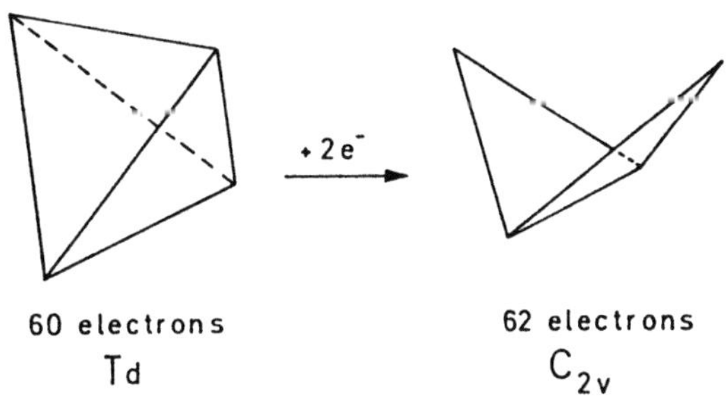

In the 60-electron species, twelve electron pairs are used for metal-ligand bonding and twelve more pairs are essentially nonbonding lone pairs on the metal vertices. This leaves six electron pairs for cluster bonding. These six pairs precisely occupy six bonding molecular orbitals of $a_1 + t_2 + e$ symmetry, which are linear combinations of six edge-localized single bonds [59, 61]. The six corresponding edge antibonding interactions give rise to six highly energetic empty MO's of $t_1 + t_2$ symmetry. Clearly upon reduction, the extra electron pair will enter the t_1 or t_2 antibonding orbitals. The triple degeneracy of these orbitals gives rise to a pronounced JT instability. It may be shown that a T type instability results, with a dominant warping term favoring the C_{2v} epikernel solution of the open butterfly shape [62]. This solution is in agreement with the epikernel principle for T terms in tetrahedral symmetry.

While 62-electron clusters of the type M_4L_{12} frequently adopt the C_{2v} *open* butterfly shape, M_4L_{16} clusters with the same electron count, usually form a D_{2h} *flat* butterfly geometry [63]. The parent electron precise cluster of the latter butterfly shape is not the 60-electron tetrahedron but the 64-electron square plane.

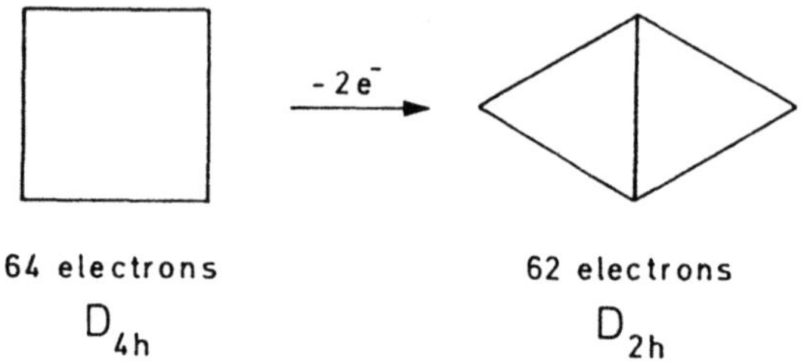

The D_{4h} geometry is exemplified by the 64-electron cluster $Pt_4(CH_3COO)_8$. In this cluster the eight bridging acetates are arranged in such a way that each Pt atom is surrounded by four oxygen atoms, forming a *cis*-divacant octahedron [64].

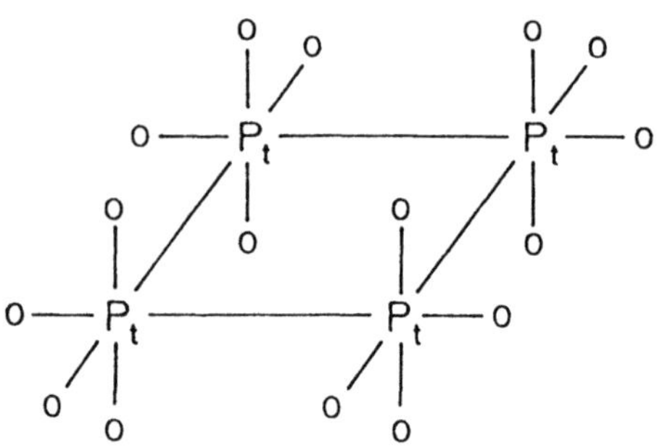

Leaving out twelve nonbonding lone pairs (occupying the t_{2g}-like *d*-orbitals on the metal vertices) and sixteen metal-ligand bonding pairs, only four pairs remain for cluster bonding. These precisely fit into the four MO's of $a_{1g} + b_{2g} + e_u$ symmetry, which describe a completely bonding square. The corresponding four antibonding MO's of $a_{2g} + b_{1g} + e_u$ symmetry remain empty. Withdrawal of one electron pair from the bonding manifold will destabilize the square planar geometry. In a one-electron view, the observed [65] flat butterfly distortion in a cluster, such as $Re_4(CO)_{16}^{2-}$, may be attributed to the JT activity of the twofold degenerate e_u orbital, the D_{2h} butterfly geometry being an epikernel of the tetragonal $e_u \times (b_{1g} + b_{2g})$ problem (Cf. Eq. 12).

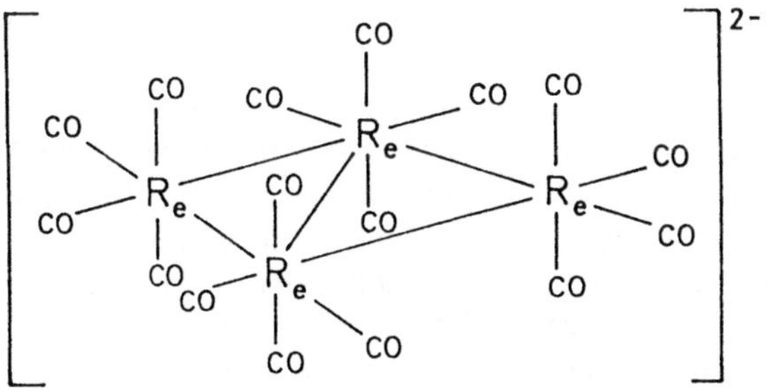

The two canonical e_u components are e_u^x and e_u^y, and comprise in plane *p*-like π-orbitals and *s*-like σ-orbitals. Clearly a $D_{2h}(C_2')$ distortion which involves compression of the x axis and elongation of the y axis will split the e_u orbitals. The energy of e_u^y will be lowered while the energy of e_u^x is being raised. Eventually the latter orbital may become

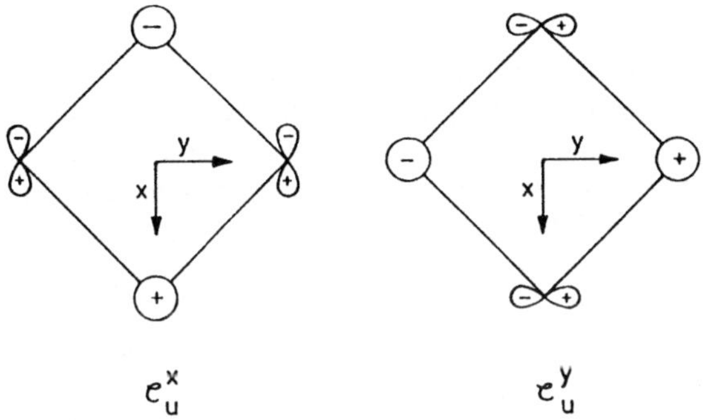

antibonding, leaving a 6-electron diamagnetic ground state, based on the $a_{1g} + b_{2g} + e_u^y$ MO's. Under $D_{2h}(C_2')$ symmetry, this butterfly configuration is labeled $(a_g)^2(b_{1g})^2(b_{2u})^2$. Interestingly the same ground state assignments may be obtained by using the more current fragment orbital approach [66]. According to this method the MO structure of the butterfly geometry is constructed [67] by considering the effect of edge-bridging a completely bonding M_3L_{12} triangle [68] with a ML_4 fragment.

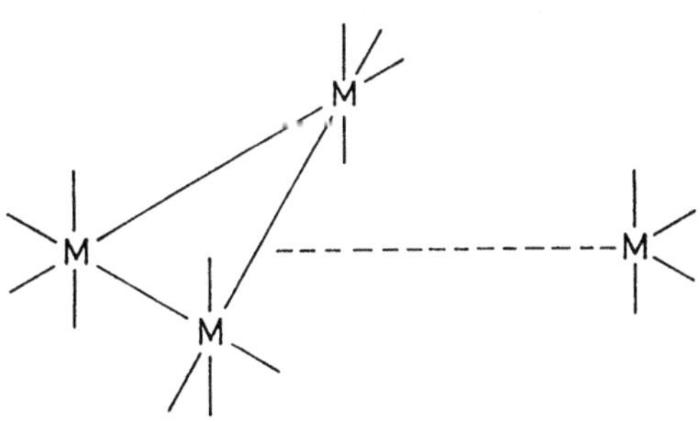

While the two approaches are complementary, the JT methodology offers the advantage that its focus is not limited to the minimum energy structures to be explained, but simultaneously envisages the essential parts of the potential energy surface, on which these structures are situated. In principle such an approach allows a unified treatment of static and dynamic effects.

Finally cluster compounds are expected to offer the first examples of JT instability in icosahedral symmetry. An interesting compound in this respect is the so called *buckminsterfullerene,* which is a C_{60} carbon cluster with a regular truncated icosahedral structure. A quantum-chemical investigation of the T × h instability in the excited states of C_{60} and the ground state of the radical anion C_{60}^{-} has already been undertaken [69]. Possibly, cations of icosahedral carbon clusters could also provide the first examples of fourfold and fivefold degenerate coupling cases. Hence the dodecahedral C_{20}^{+} cage is expected to exhibit a G × (g + h) instability [70], while the buckminsterfullerene cation, C_{60}^{+}, should exemplify the complex H × (g + 2h) coupling problem.

7 Conclusions

In this paper we have presented the epikernel principle in the framework of a unified treatment of JT potential energy surfaces. The starting point of this treatment for all JT problems considered was a simplified JT hamiltonian, containing only linear and harmonic terms, and with equal JT stabilization energies for all participating modes. This starting point is often referred to as the limit of *degenerate coupling*. Most studies on this coupling condition appeared in the physics literature [4, 27, 71–73]. Under degenerate coupling the eigenvalues of the hamiltonian will exhibit rotational symmetry [71, 72]. Accordingly the minimal energy points form a continuous equipotential trough, surrounding the JT origin. For a twofold degenerate JT instability this trough was shown to be a circular or elliptic valley (Figs. 3, 4). For a threefold degenerate JT instability a constant sphere in function space was obtained.

The great utility of such a cylindrical or spherical starting point is that deviations from the ideal coupling limit, which inevitably arise when considering realistic JT systems, may easily be described by cylindrical or spherical harmonics (Cf. Eqs. 15, 16, 19, 20). These so-called warping terms give rise to the appearance of extrema on the JT surface, such as energy sinks, local hill tops, and saddle points. The warping terms must be compatible with the structure of the underlying configuration space. This means that they must be totally symmetric under the point group symmetry of the configuration space.

In chemical applications only two types of warping effects appear to be relevant. One is due to unequal JT stabilization energies for non-degenerate modes (mode splitting). The other stems from the introduction in the JT hamiltonian of the remaining quadratic or bilinear interactions, not included in the harmonic terms. These mechanisms have been shown to give rise to dominant warping functions of the lowest possible ranks [4]. Accordingly only a limited folding of the surface is actually realized. Precisely enough extremal points are found to be available to match the epikernel sites of the configuration space. Once the wells and hills of the warping potential are anchored in these sites, as required by symmetry, no additional folds are left to produce extrema at the more numerous kernel sites. Such a warping process inevitably leads to the epikernel principle.

From this it should be clear that the epikernel principle refers to the particular symmetry breaking properties of chemically relevant JT hamiltonians. Particularly, it does *not* claim that molecules in non-degenerate states should have maximal symmetry.

Moreover, in all applications the range of the principle should explicitly be limited to the JT active modes, that are allowed under the first order JT selection rule. This limitation is necessary to derive the structure of the JT surface from the degenerate coupling limit. As such the principle fails to offer a complete description of the anomalous JT problems, comprising symmetry destroying modes, that are forbidden to first order. To our knowledge the only case in point, studied to date,is the anomalous JT effect in the degenerate half-filled shell states of certain diradicals.

In conclusion the fundamentals of the epikernel principle are seen to be a) symmetry contraints imposed on the surface by the structure of the configuration space, b) dominance of linear and quadratic vibronic coupling in the breakdown of degeneracy.

In essence a similar rationale underlies the Murrell-Laidler theorem on the structure of potential energy surfaces near transition states [74, 75]. The assertion that a chemical transition state should always be a saddle point between only two valleys – a reactant valley and a product valley – indeed is based on the assumption that the potential near the transition state is dominated by second-order interactions. Further interesting analogies can be drawn with the *maximality principle* for phase transitions in the solid state [14], the *principle of least motion* [9] in chemical reaction theory and the related *principle of maximal symmetry* expressed by Rodger and Schipper [76].

Acknowledgement. This work was supported by a research fund from the Belgian Government (Programmatie van het Wetenschapsbeleid). One of the authors (A.C.) is indebted to the Belgian National Science Foundation (NFWO) for a senior research associateship.

8 References and Notes

1. Bersuker IB (1984) The Jahn-Teller effect and vibronic interactions in modern chemistry, Plenum, New York
2. Bersuker IB (1984) The Jahn-Teller effect, a bibliographic review, Plenum, New York
3. Ceulemans A, Beyens D, Vanquickenborne LG (1984) J. Am. Chem. Soc. 106: 5824; Eq. 10 of this reference contains two sign errors. The bilinear terms in $Q_\theta Q_\zeta$ and $Q_\theta Q_\eta$ should both have a minus sign
4. Ceulemans A (1987) J. Chem. Phys. 87: 5374
5. McDowell RS (1965) J. Mol. Spectrosc. 17: 365
6. Murray-Rust P, Bürgi H-B, Dunitz JD (1979) Acta Crystallogr., Sect. A 35: 703
7. Woodward RB, Hoffmann R (1970) The conservation of orbital symmetry, Verlag Chemie, Weinheim
8. Jahn HA, Teller E (1937) Proc. R. Soc. London, Ser. A 161: 220
9. Pearson R (1976) Symmetry rules for chemical reactions, Wiley, New York
10. Bader RFW (1962) Can. J. Chem. 40: 1164
11. Halevi, EA (1975) Helv. Chim. Acta 58: 2136
12. Katriel J, Halevi EA (1975) Theor. Chim. Acta 40: 1
13. Melvin MA (1956) Rev. Mod. Phys. 28: 18
14. Ascher E (1977) J. Phys. C. 10: 1365
15. Rodger A, Schipper PE (1987) J. Phys. Chem. 91: 189
16. Fritzer HP (1979) NATO Adv. Study Inst. Ser., Ser. B 43: 179
17. Jotham RW, Kettle SFA (1971) Inorg. Chim. Acta 5: 183
18. Herzberg G (1945) Infrared and Raman spectra of polyatomic molecules, Van Nostrand, Princeton
19. Oepik U, Pryce MHL (1957) Proc. R. Soc. London, Ser. A 238: 425

20. For real representations, the matrix element $\langle \Gamma\gamma_i | \partial\mathcal{H}/\partial Q_{\Lambda\lambda} | \Gamma\gamma_j \rangle$ will be zero, unless $\Lambda = \Gamma \times \Gamma$. Furthermore, since the $\mathbb{H}$ matrix is real and hermitian, Λ must be contained in the symmetrized square of Γ. According to group theory, $[\Gamma]^2$ contains the totally symmetric representation exactly once. Since A_1 vibrations cannot lower the symmetry of the JT origin, the JT active vibrations will be contained in $[\Gamma]^2 - A_1$, as expressed in Eq. (1).
21. Liehr AD (1963) Progr. Inorg. Chem. 5: 385
22. Bersuker IB, Polinger VZ (1982) Adv. Quantum Chem. 15: 85
23. The projection is bijective since two antipodal points on the globe, (a_x, a_y, a_z) and $(-a_x, -a_y, -a_z)$, refer to the same eigenvalue. Hence the plot can be restricted to one hemisphere.
24. Khlopin VP, Polinger VZ, Bersuker IB (1978) Theor. Chim. Acta 48: 87
25. Butler PH (1981) Point group symmetry applications, Plenum, New York. The function given in this reference has been transformed to the more convenient coordinate system in: Boyle LL, Parker YM (1980) Molec. Phys. 39: 95
26. Interestingly similar drawings may be encountered in the study of rotational energy surfaces of cubic and icosahedral molecules. See: Harter WG (1984) Journal of Statistical Physics 36: 749; Harter WG, Weeks DE (1986) Chem. Phys. Lett. 132: 387
27. O'Brien MCM (1969) Phys. Rev. 187: 407
28. Bersuker IB, Polinger VZ (1974) Sov. Phys.–JETP (Engl. Transl.) 39: 1023
29. Bacci M, Ranfagni A, Fontana MP, Viliani G (1975) Phys. Rev. B: Solid State 11: 3052
30. Kataoka M, Nakajima, T (1984) Theor. Chim. Acta 66: 121
31. Nakajima T, Kataoka, M (1984) Theor. Chim. Acta 66: 133
32. Ammeter JH, Zoller L, Bachmann J, Baltzer P, Gamp E, Bucher R, Deiss E (1981) Helv. Chim. Acta 64: 1063
33. Zoller, L, Moser E, Ammeter JH (1986) J. Phys. Chem. 90: 6632
34. Borden WT, Davidson ER, Feller D (1981) J. Am. Chem. Soc. 103: 5725
35. Knight LB Jr, Steadman J, Feller D, Davidson ER (1984) J. Am. Chem. Soc. 106: 3700
36. Paddon-Row MN, Fox DJ, Pople JA, Houk KN, Pratt DW (1985) J. Am. Chem. Soc. 107: 7696
37. Takeshita K (1987) J. Chem. Phys. 86: 329
38. Caballol R, Català JA, Problet JM (1986) Chem. Phys. Lett. 130: 278
39. Walther BW, Williams F: J. Chem. Soc. Chem. Comm. 1982: 270
40. Davidson ER, Borden WT (1977) J. Chem. Phys. 67: 2191
41. Borden WT, Davidson ER, Feller D (1980) J. Am. Chem. Soc. 102: 5302
42. Ceulemans A, Beyens D, Vanquickenborne LG (1982) J. Am. Chem. Soc. 104: 2988. For a rigorous proof of Jahn-Teller inactivity of half-filled shell states, see: Ceulemans A (1985) Meded. K. Acad. Wet., Lett. Schone Kunsten Belg., Kl. Wet. 46: 82
43. Deeth, RJ, Hitchman MA (1986) Inorg. Chem. 25: 1225
44. Reinen D, Friebel C (1979) Struct. Bonding 37: 1
45. Riley MJ, Hitchman MA, Reinen D (1986) Chem. Phys. 102: 11
46. Hathaway BJ (1984) Struct. Bonding 57: 55
47. Bacci M (1979) Chem. Phys. 40: 237
48. Reinen D, Allmann R, Baum G, Jakob B, Kashuba U, Massa W, Miller GJ (1987) Z. anorg. allg. Chem. 548: 7
49. Chrichton O, Poliakoff M, Rest AJ, Turner JJ: J. Chem. Soc., Dalton Trans. 1973: 1321
50. Bacci M (1978) Chem. Phys. Lett. 58: 537
51. Reinen D, Atanasov M, Nikolov GSt, Steffens F (1988) Inorg. Chem. 27: 1678
52. Parrot R, Naud C, Gendron F, Porte C, Boulanger D (1987) J. Chem. Phys. 87: 1463
53. Nieke C, Reinhold J (1984) Theor. Chim. Acta 65: 99
54. Breza M, Pelikán P, Boča R (1986) Polyhedron 5: 1607
55. The plasticity effect is a solid state effect which refers to changes of the direction of distortion in the coordination sphere of JT active metal ions, as a result of changes in the crystal environment. The effect points to the absence of substantial barriers in the trough potential.
56. Poliakoff M, Ceulemans A (1984) J. Am. Chem. Soc. 106: 50
57. Wade K: J. Chem. Soc. Chem. Comm. 1971: 792
58. Mingos DMP (1972) Nature 236: 99
59. Wade K (1980) In: Johnson BFG, (ed), Transition metal clusters, Wiley, Chichester, chap III
60. Johnson BFG, Benfield, RE (1981) Top Stereochem. 12: 253
61. Ceulemans A, Fowler PW (1985) Inorg. Chim Acta 105: 75

62. Ceulemans A (1986) J. Chem. Phys. 84: 6442
63. Lauher JW (1978) J. Am. Chem. Soc. 100: 5305
64. Carrondo MAAF de CT, Shapski AC (1978) Acta Crystallogr. Sect. B 34: 1857
65. Churchill MR, Bau R (1986) Inorg. Chem. 7: 2606
66. Hoffmann R (1982) Angew. Chem. Int. Ed. 21: 711
67. Evans DG, Mingos DMP (1983) Organometallics 2: 435. The D_{2h} symmetry labels in Fig. 4 of this article are erroneous, and should read b_{1g}, b_{2u}, a_g, instead of b_{2u}, $1a_{1g}$ and $2a_{1g}$. These corrected assignments coincide with the results from the JT treatment. Private communication from D. G. Evans.
68. Mealli C (1985) J. Am. Chem. Soc. 107: 2245
69. Negri F, Orlandi G, Zerbetto F (1988) Chem. Phys. Lett. 144: 31
70. A general analysis of the linear $G \times (g + h)$ problem has been carried out recently. The results fully confirm the epikernel principle. Ceulemans A, Fowler PW (1989) Phys. Rev. A 39: 481
71. Pooler DR (1978) J. Phys. A 11: 1045
72. Pooler DR (1980) J. Phys. C 13: 1029
73. Judd BR (1974) Can. J. Phys. 52: 999
74. Murrell JN, Laidler KJ (1968) Trans. Farad. Soc. 64: 371
75. Stanton, RE, McIver JW (1975) J. Am. Soc. 97: 3632
76. Rodger A, Schipper PE (1986) Chem. Phys. 107: 329

The Prediction and Interpretation of Bond Lengths in Crystals

M. O'Keeffe

Department of Chemistry, Arizona State University, Tempe AZ 85287, USA

The concept of bond valence and its correlation with bond length are reviewed. It is shown how to augment bond valence sums at an atom with other constraints so that bond lengths can be predicted for a given topology. These constraints can be read directly from the connectivity matrix for the structure. The physical reasons for the constraints are analysed, and this analysis leads in turn to a reformulation of Pauling's electrostatic valence sum rule. The apparent valences of atoms, calculated as a sum of bond valences derived from bond lengths, are often significantly different from the actual valences. It is demonstrated that such observations are often diagnostic of non-bonded repulsions. Limitations and possible extensions of the method are outlined.

1 Introduction 162
2 Atom and Ion Radii and Bond Lengths 162
3 The Bond Valence Method 163
4 Bond Valences in Molecules 164
5 Application to Crystals 166
6 Bond Length Prediction in Crystals 169
7 Justification of the Ring and Circuit Procedures and Further Examples 175
8 Pauling's Electrostatic Valence Sum Rule 178
9 The Olivine and "Inverse Spinel" Structures 179
10 Apparent Valences 180
11 Limitations to Predictions of Bond Lengths 185
12 "Unusual" Valences 186
13 Concluding Remarks 186
14 Appendix 1. The Bond Valence Method and Kirchhoff's Laws 187
15 Appendix 2. Bond Valence Parameters 189
16 References 189

© Springer-Verlag Berlin Heidelberg 1989

1 Introduction

This article is concerned with two aspects of interatomic distances in crystals. The first problem is that of predicting interatomic distances to be expected in a given crystal structure as yet unknown. The second concerns the interpretation of observed distances in known structures. As they rely largely on empirical correlations, these two topics are closely interwoven. Most of the discussion will be devoted to materials such as oxides, nitrides and fluorides for which a large data base of structures exists, but in fact the treatment is of wider applicability.

Historically, bonding interatomic distances in crystals were first discussed in terms of atomic or ionic radii; however the concept of bond valence and its correlation with bond length has now come to the fore, largely because of the work of W. H. Zachariasen and of I. D. Brown. A brief summary of the reasons for preferring the latter approach is given first. In fact a case will be made for abandoning the concept of ionic (or atomic) radii as currently applied in that area of crystal chemistry (oxides and similar materials) that is the main concern here.

In the discussion of the two problems alluded to above it is found convenient to distinguish two quantities which are called *bond valence* and *apparent bond valence.* Bond valence, as the term is used and explained below, is essentially a topological property of a crystal structure. The apparent bond valence is a property of a bond that is calculated from its length using a predetermined equation, and in this sense is an experimentally determined quantity. The failure to distinguish between these two quantities has been the source of some confusion in the past.

In the discussion of the bond valence method two important problems are skirted. The first, currently a topic of lively debate, is the problem of deciding whether two atoms that are in close proximity in a structure are in fact "bonded" together. The second is the problem of determining, *a priori,* the valence of an atom. These problems are circumvented by avoiding situations where their solution is not "obvious". The treatment of the bond valence method owes much to the work of Brown [1, 2] but also contains some new results. It complements an earlier account [3] of new methods of *describing* structures. Historically the concept of bond valence derives from the Pauling [4] *bond number* as applied to metals and intermetallic compounds and subsequently applied to oxides by Bystrom and Wilhelmi [5].

2 Atom and Ion Radii and Bond Lengths

In the early days of crystallography, when relatively few bond lengths were known, and those not very accurately, it was observed [6] that bond lengths were approximately a sum of atomic radii. This idea was further developed by Slater [7] and a first approximation to bond lengths may be obtained as a sum of Bragg-Slater atomic radii. This approach does not however, allow for any variation of bond lengths between two given types of atom.

The difficulty of more exact treatments led to the widespread adoption of the ionic model of bonding, and bond lengths in crystals have been more commonly discussed in terms of ionic radii. However, the actual implementation of this model was unfortunate for several reasons. Thus it was widely assumed that in oxides (to take a specific example), oxygen had a fixed radius – usually chosen in the range of 1.32–1.40 Å – although the most refined treatments, such as that of Shannon and Prewitt [8] allowed some small variability in the size of oxygen. This size for oxygen led to sizes for cations that were much too small and to the ascription of essentially all the variability of bond lengths to variations in cation radius. Atomic wave functions (for the lighter atoms and positive ions at least) have been known for some time and it is known that core and valence regions are rather well defined for an atom such as Si (again to take a specific example). Accordingly, the radius of Si^{4+} (defined in a rigorous way as the radius of a sphere centered at the Si nucleus and containing a net charge of + 4) is fairly constant (and is about 0.61 Å – O'Keeffe, to be published). As the core electrons are very tightly bound, this core radius is not subject to significant variation when a Si atom is placed in different environments in crystals; in particular the radius thus defined will not vary with coordination number.

Apart from the unphysical nature of ionic radii, they are subject to the further criticism that they do not lead to correct predictions in mixed anion compounds such as oxyfluorides [9]. Also they do not allow one to handle correctly compounds in which there is a range of bond lengths within a given coordination polyhedron. One example of the latter is afforded by the case of B_2O_3 which is discussed below.

The difficulties associated with the use of ionic radii to predict bond lengths in crystals are largely overcome in the bond valence method, and we now turn to this topic.

3 The Bond Valence Method

Consider an atom A forming Z equivalent (i.e. symmetry related) bonds to atoms X. The bond valence, ν_{AX} in this case is just the valence of A, V_A, divided by the number of bonds.

$$\nu_{AX} = V_A/Z \tag{1}$$

Thus in MgO in which Mg is bonded to six equidistant oxygen atoms at the vertices of a regular octahedron the bond valence $\nu_{MgO} = 2/6 = 1/3$. In this special (symmetrical) case the bond valence, as given by Eq. (1), is the same as the Pauling [10] electrostatic bond strength[1] but these two concepts should be carefully distinguished.

An important generalization [13] is to cases where there are non-equivalent bonds from an atom i to atoms j. In such a case bond valences are *defined* to be such that

$$\Sigma_j \nu_{ij} = V_i \tag{2}$$

1 Many authors (including this one, previously) use the term *bond strength* for what we now prefer to call *bond valence* after Brown [1, 2] to avoid confusion with the Pauling electrostatic bond strength. The former term also suggests a relationship to *bond energy* that is not necessarily to be assumed, although bond strength is sometimes used to mean bond dissociation energy [11] and empirical (non-linear) correlations of bond energy with bond valences have been used with some success in approximating energy surfaces appropriate for chemical reactions [12]

This equation does not suffice to determine the individual bond valences in a general low symmetry environment. Zachariasen proposed using empirical correlations of bond lengths with bond valences to determine individual bond valences appearing in Eq. (2).

In what follows, two separate problems that should be clearly distinguished are discussed. The first is to find a method of determining the individual bond valences appearing in Eq. (2) – a partial solution to the problem which is of practical utility in many instances is presented. The second problem is the determination of bond valences from experimental data (bond lengths). As bond lengths are affected by factors other than bond valence, bond valences determined in this way are referred to as *apparent bond valences* and symbolized v'_{ij}. The sum of apparent bond valences for all the bonds to a given atom are called the *apparent valence*, V'_i.

$$\Sigma_j v'_{ij} = V'_i \tag{3}$$

A simple example may be of use in distinguishing between V and V' and between v and v'. In both CaO and $CaCO_3$ (calcite), Ca forms six equivalent bonds to oxygen, so that $V_{Ca} = 2$ and $v_{CaO} = 1/3$ in both cases; however the Ca–O bonds are of significantly different lengths in the two compounds so that if one uses a unique bond length – bond valence correlation for Ca–O bonds the apparent bond valence and the apparent valence of Ca will be different in each case.

The correlation of bond length with bond valence has been reviewed by Brown [2]. Because of its simplicity, I use:

$$v_{ij} = \exp[(R_0 - d_{ij})/b] \tag{4}$$

where R_0 is a parameter characteristic of the atom pair i, j and b is a second parameter. Usually the parameters in Eq. (4) are fitted by least squares to give the best values in accord with Eq. (2) for a number of well-determined structures. In practice b can be maintained constant at $b = 0.37$ Å for a wide variety of bonds [14]. In what follows I use this value of b and the values of R_0 recommended by Brown and Altermatt where available. It is important to use Eq. (4) only for interpolation within the range of bond lengths and valences observed in the structures used to determine the parameters – extrapolation either to very long or very short bonds may well lead to incorrect apparent valences.

4 Bond Valences in Molecules

Simple molecules are considered first as they illustrate many of the problems arising in crystals. A noncyclic molecule (for example formaldehyde, H_2CO) with N atoms will contain $N - 1$ bonds. There will be N equations of the type of Eq. (2), one for each atom. Of these equations, only $N - 1$ are independent because of the overall valence constraint (electrical neutrality in an ionic model). Thus the $N - 1$ bond valences are uniquely determined in an almost trivial way. ($v_{CH} = 1$, $v_{CO} = 2$ and $V_C = V_O + 2\ V_H$ in this example.)

Now consider cyclic molecules. For each bond closing a ring a new bond valence is introduced without adding a new bond valence sum. If there were *n different kinds* of ring, *n* extra conditions would be needed in order that the bond valences be determined. Consider the simple ring *A–X–B–X*:[2]

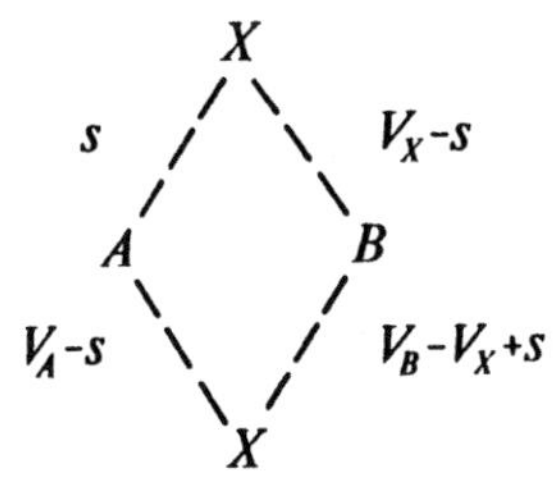

where the bond valences satisfying the bond valence sums are indicated in terms of an undetermined parameter *s*. To solve this particular problem appeal is made to the probability of there being a horizontal mirror in the above molecule so that the valences of the two *A–X* bonds would be equal, in which case $s = V_A/2$. As $V_A + V_B = 2\ V_X$ one has the "obvious" solution:

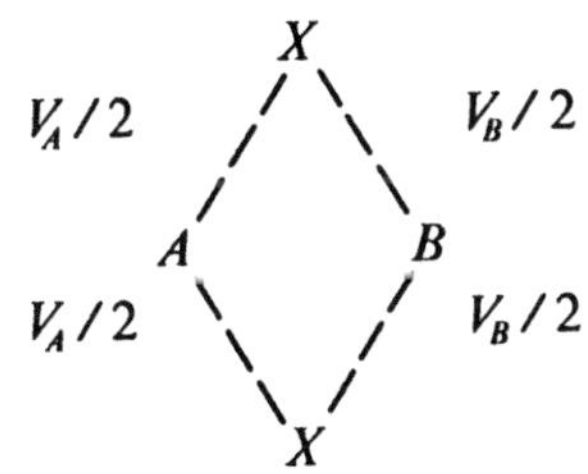

Now consider a more general case of a ring *A–X–B–Y* (such as the molecule LiOBeF). The situation is again:

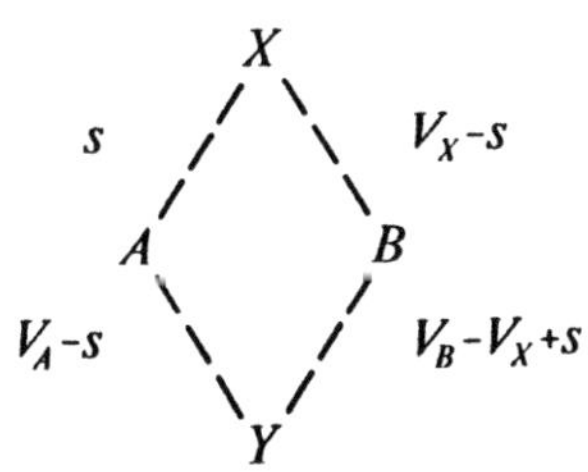

To solve this problem a hint is taken from what was done for the first ring. In that case the difference between the valences of the two *B–X* bonds (and of the two *A–X* bonds)

2 The convention is used that electropositive elements ("cations") *A, B, C*... are bonded only to electronegative elements ("anions")... *X, Y, Z*. Only compounds in which the distinction is clear are considered

was made as small as possible (zero). In the present case $\delta_A = v_{AX} - v_{AY} = 2\,s - V_A$ and $\delta_B = v_{BX} - v_{BY} = 2\,V_X - V_B - 2\,s$. The sum $\delta_A + \delta_B = 2\,V_X - V_B - V_A$ (a constant), but the most "nearly symmetrical" situation would be for $\delta_A = \delta_B$. If this condition is accepted, and it is recalled that $V_A + V_B = V_X + V_Y$, the solution is found to be:

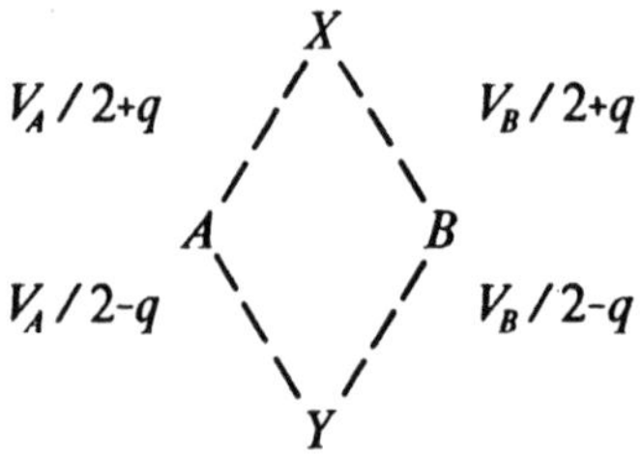

where $q = (V_X - V_Y)/2$.

There is an interesting way to interpret this procedure that was suggested by Brown [15]. Bonds are treated as vector quantities directed from "cations" to "anions" and the bond valences summed around the ring. Then starting from A and proceeding clockwise:

$$v_{AX} - v_{BX} + v_{BY} - v_{AY} = \delta_A - \delta_B = 0 \tag{5}$$

Thus the sum of the bond valences (considered as having a sign) around the ring is zero.

As each time a new type of ring is made in a molecule, one new bond strength equation is required, and as this is provided by the ring sum, it can be seen that the problem is exactly determined in the sense that there are as many independent sums as there are different bond valences to be determined. This procedure is in some ways analogous (but see the appendix to this paper) to the solution of problems in electrical networks using Kirchhoff's junction and loop laws [15, 16]. In what follows it is shown how the procedure can be simply used in practice to predict bond lengths in crystals and consequently to provide at least partial justification for it.

5 Application to Crystals

Exactly the same procedure as outlined above is used in applications to crystals. Bonds between pairs of different atoms are considered to have separate valences to be determined by valence sums. Atoms are considered "different" if they are distinct in the crystallographic sense. A difficulty that often arises in practice is that bonds between pairs of atoms that are related by symmetry are not necessarily themselves related by symmetry: a familiar example is that of the corundum (α-Al_2O_3) structure in which there is only one kind of Al atom and one kind of O atom but two different Al–O bonds that are of unequal lengths. The method of analysis used above would ascribe equal bond valences to these two bonds, but, because they are of unequal lengths, different apparent bond valences would be ascribed to them.

Rings of four atoms will give no new information if they contain less than four different kinds of atom (see the first ring considered in the last section). Rings of six[3] that contain more than a pair of identical atoms will likewise contain no useful bond valence sums. A ring of six in which there is a pair of equivalent atoms, such as *A–X–B–Y–C–Y* is equivalent (in this context) to a four membered ring *A–X–B–Y*. A ring of six atoms, all different, such as *A–X–B–Y–C–Z* generates a new type of equation:

$$v_{AX} - v_{BX} + v_{BY} - v_{CY} + v_{CZ} - v_{AZ} = 0 \tag{6}$$

and similarly for larger rings if they need to be identified.

One does not actually have to identify the rings of atoms in the structure – an algorithm is given here that produces exactly equivalent results. The connectivity matrix is first set out as in the example below in which an asterisk indicates a bond between a pair of atoms:

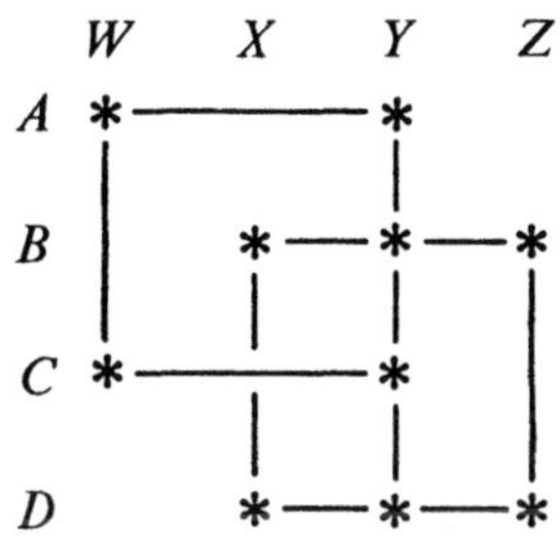

thus, in this example *A* is bonded to *W* and *Y* and so there are asterisks in the positions *AW* and *AY*, etc. There are eight kinds of atom and seven independent bond valence sums at the atoms. (Combination of any seven with the eighth will merely yield the constraint on stoichiometry $V_A + V_B + V_C + V_D = V_W + V_X + V_Y + V_Z$.). On the other hand there are ten kinds of bond (represented by *). One must therefore find three circuits and these are outlined on the diagram. Distinction is made between a *ring* of atoms that actually exists in a crystal structure and a *circuit* on a connectivity matrix which may not in general correspond to an actual ring of atoms but which will always yield constraints on bond valences equivalent to those obtained by summing around rings.

The top left circuit in the diagram above connects the points *AW, AY, CY, CW*. (Note that points such as *BY* are not considered part of this circuit as it is not on a corner of it. The reason for this is that horizontal lines correspond to "cations" (*A, B, C,* etc.) and vertical lines to anions, so that horizontal and vertical lines must alternate on a circuit.) This circuit yields the equation:

$$v_{AW} - v_{AY} + v_{CY} - v_{CW} = 0$$

Two other circuits in the figure yield the equations

3 It should be clear that in the type of crystal being considered, all rings of bonds must consist of an even number of atoms

$$\nu_{BX} - \nu_{BY} + \nu_{DY} - \nu_{DX} = 0$$

$$\nu_{BY} - \nu_{BZ} + \nu_{DZ} - \nu_{DY} = 0$$

It may be verified that the other four-membered circuit in the figure, connecting *BX, BZ, DZ, DX,* yields an equation that is just the sum of the last two. Note that although the circuits identified on the connectivity matrix may not exist in the actual crystal structure; if they do, they are dual to them in the sense that points on the connectivity matrix diagram correspond to bonds in the crystal structure and that the lines on the matrix diagram correspond to atoms in the crystal structure.

A second example is:

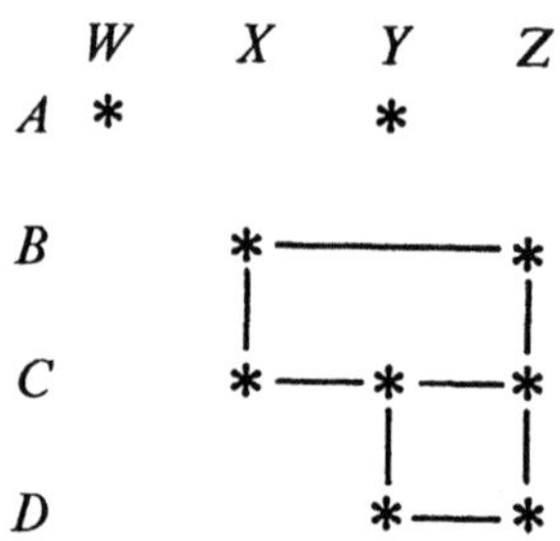

Now there are nine bonds and again seven independent bond valence sums at the atoms, so that two circuits must be identified. There is a four-membered circuit connecting *CY, CZ, DZ, DY,* and a second connects *BX, BZ, CZ, CX*. There is also a six-membered circuit connecting *BX, BZ, DZ, DY, CY, CX*. The corresponding equation obtained from this circuit is:

$$\nu_{BX} - \nu_{BZ} + \nu_{DZ} - \nu_{DY} + \nu_{CY} - \nu_{CX} = 0$$

which is just the sum of the two four-membered circuits. Note again that the point *CZ* was omitted as it is not at the corner of the circuit. The valence of the *A–W* bond is determined directly by the bond valence sum at *W*, as *W* is bonded only to *A*, and that the valence of the *A–Y* bond is then determined from the bond valence sum at *Y* (which entails only ν_{AW} and ν_{AY}).

A third example is more complex and is that derived from the structure of β-Mg_2SiO_4 [17] which may be written Mg(1)Mg(2)Mg(3)$_3$Si$_2$O(1)O(2)O(3)$_2$O(4)$_4$. The connectivity matrix is now set out using Greek letters for the bond valences

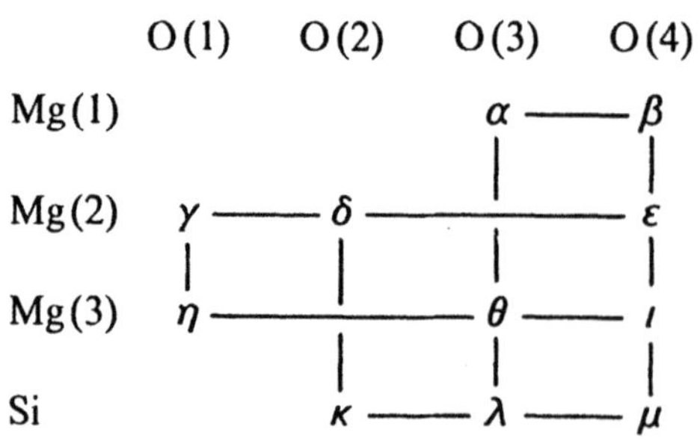

There are eleven bond valences to determine and seven independent bond valence sums at the atoms. Accordingly four circuit sums must be identified. Four independent[4] four-membered circuits can be seen: αβμλ, δεμϰ, θιμλ and γειη. There is also a six-membered one δεβαλϰ but this has no new sides and thus contains no new information – this can be verified as follows. From the first two four-membered circuits:

$$\alpha - \beta + \mu - \lambda = \delta - \varepsilon + \mu - \varkappa = 0$$

From the six-membered circuit:

$$\delta - \varepsilon + \beta - a + \lambda - \varkappa = 0$$

which is readily derived from the two equations above.

If there are more than two kinds of cation and more than two kinds of anion, it will not be always possible to express the circuit sums in terms of sums around four-membered circuits, as the following simple counter-example shows:

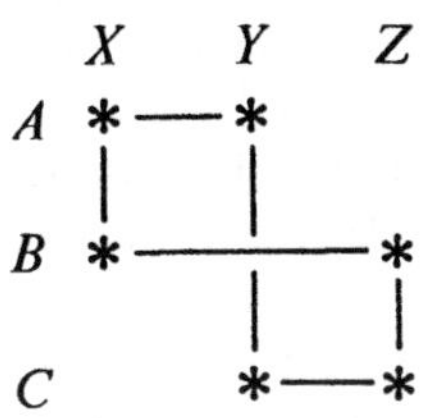

6 Bond Length Prediction in Crystals

There is currently great interest in predicting crystal structures. In principle it can be done with high precision quantum mechanical calculations, but in practice this is only possible for the simplest of structures. One then resorts to more empirical methods. These are not of concern here, other than to point out that in many instances if the bond lengths were known, the structure would either be known or at least severely constrained.

Here the application of the bond valence method to the bond length problem is discussed. One needs first a set of parameters R_0 and b for each kind of atom pair. Many of these have been provided by Brown [2] and by Brown and Altermatt [14] and reference should be made to those papers (and the papers they cite) for the techniques of utilizing the existing structural data base for this purpose. In some instances it may be useful to develop an *ad hoc* set of parameters for the analysis of a series of related structures.

The simplest case to consider is that of structures with only one parameter and in which there is just one bond length. These structures are of necessity cubic (lattice

4 In this context a new circuit is independent if it yields an equation that is not a linear combination of equations derived from circuits previously identified

parameter *a*) and have atoms located at special positions at the intersections of symmetry elements. Familiar examples include the **NaCl** (B1), **Sphalerite** (B2), **Fluorite** (C1) and **Cuprite** (C3) structures[5]. Other simple binary structures involve more than one parameter and require some further constraints if the structure is to be fully determined. For example the **wurtzite** structure of ZnO has three parameters and two independent bond lengths in the ZnO_4 (or OZn_4) tetrahedra. One could require that the tetrahedra be regular and thus get a first approximation to the structure; a better approximation would be obtained if one had some insight into why in fact the two bond lengths are different [18].

Other simple cases where the number of parameters is equal to the number of independent bond lengths are the **spinel** and **garnet** structures [3], which are completely determined if the bond lengths are known. They will not detain us here, the present goal being rather to determine expected bond lengths from bond valences in more complex structures.

There is an important class of structures in which there are only three kinds of atom. Here there will be two independent bond valence sums and only two different bond valences. An example is the structure of B_2O_3(II) which is written[6] $^{iv}B_2{}^{iii}O(1)_2{}^{ii}O(2)$ – emphasizing the ternary nature of the structure (Si_2N_2O is isostructural). Examining the bond valence sums at the anions it may be seen immediately that $v[B–O(1)] = 2/3$ and $v[B–O(2)] = 2/2 = 1$. The coordination of B is $\{B\}O(1)_3O(2)$, so that the BO_4 tetrahedron has three bonds with $v = 2/3$ and one bond with $v = 1$. The bond lengths calculated from the Brown and Altermatt parameter ($R_0 = 1.371$ Å) are 1.52 (3 ×) and 1.37 Å. The observed values are [8] 1.512, 1.507, 1.506 and 1.373 Å. This example illustrates several important points: most of the distortions of the BO_4 tetrahedron are accounted for as a consequence of the constraints of bond valence sums. On the other hand the bond valence method will not account for the (small) differences in the three independent B–O(1) distance as the three bonds are all assumed to have the same valence. But notice that there is already a very significant improvement over the method of summing ionic radii for predicting bond lengths.

Other simple oxides (with some isostructural ternary examples in parentheses) with two kinds of anion are $^{vii}Zr{}^{iv}O(1){}^{iii}O(2)$ (ScOF, TaNO) and $^{vii}La_2{}^{vi}O(1){}^{iv}O(2)_2$ (La_2SO_2 ect.). The La_2O_3 structure has been discussed before [3] but it is returned to below. For LaOF see O'Keeffe [9].

At the next level of complexity, there are two kinds of cation and two kinds of anion, i.e. $A_aB_bX_xY_y$. Writing $v(A–X) = \alpha$, $v(A–Y) = \beta$, $v(B–X) = \gamma$ and $v(B–Y) = \delta$, the connectivity matrix is:

$$\begin{array}{ccc} & X & Y \\ A & \alpha & \!\!-\!\!\!-\, \beta \\ & | & | \\ B & \gamma & \!\!-\!\!\!-\, \delta \end{array}$$

5 Chemical formulae or mineral names in bold face refer to structure types rather than specific compounds

6 The coordinations are written as Roman superscript prefixes to the atom symbols in the compound formula. Also, to distinguish between formulas for compounds and formulas for coordination figures, in the latter case the symbol for the coordinated atom is enclosed in braces

where now the symbols for the bond valence are used to indicate a bond. The circuit equation is clearly:

$$\alpha - \beta + \delta - \gamma = 0$$

A simple example is provided by the structure of α-LiSiNO [19] – KGeNO is isostructural – that is a superstructure of the **wurtzite** type. The coordinations, and corresponding bond valence sums at the atoms for the general case *ABXY* are:

(i) $\{A\}XY_3$ $\quad \alpha + 3\beta = V_A$

(ii) $\{B\}X_3Y$ $\quad 3\gamma + \delta = V_B$

(iii) $\{X\}AB_3$ $\quad \alpha + 3\gamma = V_X$

(iv) $\{Y\}A_3B$ $\quad 3\beta + \gamma = V_Y$

It may readily be confirmed that there are only three independent equations here [e.g. (i) + (ii) = (iii) + (iv)]. These combined with the ring sum give:

$$\alpha = (4\,V_X - 3\,V_B + V_A)/8$$

$$\beta = (7\,V_A - 4\,V_X + 3\,V_B)/24$$

$$\gamma = (4\,V_X + 3\,V_B - V_A)/24$$

$$\delta = (5\,V_B + V_A - 4\,V_X)/8$$

Hence for A = Li, B = Si, X = N and Y = 0:

$$\alpha = 1/8,\ \beta = 7/24,\ \gamma = 23/24,\ \delta = 9/8$$

Detailed comparison with observed crystal structures requires both accurate structures and the development of parameters appropriate for bonds to nitrides, neither of which are generally available at present. However the important point is to recognize that the expected bond valences predict bond lengths which are different from those predicted by the "ionic radius" approach. Thus using ionic radii for four-coordinated Li and O the predicted Li–O bond length would be that for valence 1/4 rather than 7/24, etc.

There is a second simple simple structure *ABXY* that is a superstructure of the **wurtzite** structure. In this structure ($LiGaO_2$ [20] is an example) the coordinations and bond valence sums are:

(i) $\{A\}X_2Y_2$ $\quad 2\alpha + 2\beta = V_A$

(ii) $\{B\}X_2Y_2$ $\quad 2\gamma + 2\delta = V_B$

(iii) $\{X\}A_2B_2$ $\quad 2\alpha + 2\gamma = V_X$

(iv) $\{Y\}A_2B_2$ $\quad 2\beta + 2\delta = V_Y$

and hence:

$\alpha = (2\ V_X + V_A - V_B)/8$

$\beta = (3\ V_A - 2\ V_X + V_B)/8$

$\gamma = (2\ V_X - V_A + V_B)/8$

$\delta = (3\ V_B + V_A - 2\ V_X)/8$

The reader may verify that if $V_X = V_Y$ (as in $NaFeO_2$) $\alpha = \beta = V_A/4$ and $\gamma = \delta = V_B/4$.

A third example of a **wurtzite** superstructure that is of interest is the structure type A_2BX_2Y with symmetry $Cmc2_1$. Compounds with this structure include Li_2SiO_3 [21] and Li_2GeO_3 [22] in which X and Y are both O, but in which there are two differently bonded O atoms, and $LiSi_2N_3$, $LiGe_2N_3$ and $NaGe_2N_3$. Note that in the oxides Si, Ge = B but in the nitrides Si, Ge = A. Other members of this family are B_2O_3 (II) and Si_2N_2O (in which $V_B = 0$ and, as valences cannot be negative, $v_{BX} = v_{BY} = 0$). Proceeding as before, again with $v_{AX} = \alpha$, $v_{AY} = \beta$, $v_{BX} = \gamma$, $v_{BY} = \delta$ the sums are:

(i) $\{A\}X_3Y \quad 3\alpha + \beta = V_A$

(ii) $\{B\}X_2Y_2 \quad 2\gamma + 2\delta = V_B$

(iii) $\{X\}A_3B \quad 3\alpha + \gamma = V_X$

(iv) $\{Y\}A_2B_2 \quad 2\beta + 2\delta = V_Y$

Using again the ring condition $\alpha - \beta + \delta - \gamma = 0$ (the structure actually contains six-membered rings such as A–X–A–X–B–Y from which the preceding equation may be derived) and hence:

$\alpha = (2\ V_A - V_B + 4\ V_X)/20$

$\beta = (14\ V_A + 3\ V_B - 12\ V_X)/20$

$\gamma = (8\ V_X - 6\ V_A + 3\ V_B)/20$

$\delta = (6\ V_A + 7\ V_B - 8\ V_X)/20$

For oxides such as Li_2SiO_3, $\alpha = 0.3$, $b = 0.1$, $\gamma = 1.1$ and $\delta = 0.9$. Table 1 compares bond lengths in Li_2SiO_3 and Li_2GeO_3 calculated from these bond valences (using the Brown and Altermatt parameters) with the observed values. The agreement is generally fairly good – in particular it is satisfying that the irregular coordination of Si and Ge by O is well reproduced. The Li–O bond lengths are not so well accounted for, but this is to be expected as bonds of lower valence (especially β) are weaker and thus more susceptible to variations of length due to other perturbing factors[7]. To keep the comparison in

7 One possible solution to this problem [23] is simply to constrain the valences of weak bonds to be equal to their average. A better method, developed subsequently to the completion of this manuscript, is to include weights in the difference equations derived from rings according to the average bond valences s_i from a given cation (this is in fact the Pauling bond strength discussed below). Thus in this example $s_{Li} = 1/4$, $s_{Si} = 1$ and the ring equation becomes $(\alpha - \beta)/s_{Li} = (\gamma - \delta)/s_{Si}$. The resulting bond lengths are now in better accord with experiment (Table 1). It might be noted that the use of weights in this way generally makes very little difference (<0.01 Å) to predicted bond lengths, except in those cases (such as the present one) where there are wide ranges of valences in a given structure

Table 1. Calculated and observed bond lengths (Å) in A_2BX_2Y compounds

	Observed	Calculated[a]	Calculated[b]
Li_2SiO_3			
Si–O(1)	1.59	1.58	1.56
Si–O(2)	1.68, 1.69	1.66	1.69
Li–O(1)	1.93, 1.94, 1.96	1.91	1.95
Li–O(2)	2.18	2.32	2.10
Li_2GeO_3			
Ge–O(1)	1.71	1.71	1.69
Ge–O(2)	1.81, 1.81	1.79	1.81
Li–O(1)	1.92, 1.94, 1.95	1.91	1.95
Li–O(2)	2.17	2.32	2.10

[a] calculated as described in the text
[b] calculated with the modification suggested in footnote 7

perspective it should be noted that sums of "ionic" (or "atomic") radii would predict all bonds between a given pair of atoms (such as Si and O) to be of equal length.

In the above example not all bonds of (for example) the type *A–X* were related by symmetry. There are however many examples where bonds of a given type *are* related by symmetry. The example of the structure of $SrBe_3O_4$ [24] is given to further illustrate the method. There are now five kinds of atom: $SrBe(1)_2Be(2)O(1)O(2)_3$, but only five kinds of bonds as Be(2) is bonded only to O(2) (bond valence ε). The other bond valences are written α, β, γ, δ for Sr–O(1), Sr–O(2), Be(1)–O(1) and Be(1)–O(2) respectively and have the following coordinations and sums:

(i)	$\{Sr\}O(1)_3O(2)_6$	$3\alpha + 6\beta = 2$
(ii)	$\{Be(1)\}O(1)O(2)_3$	$\gamma + 3\delta = 2$
(iii)	$\{Be(2)\}O(2)_3$	$3\varepsilon = 2$
(iv)	$\{O(1)\}Sr_3Be(1)_2$	$3\alpha + 2\gamma = 2$
(v)	$\{O(2)\}Sr_2Be(1)_2Be(2)$ 2	$2\beta + 2\delta + \varepsilon = 2$

Again the ring sum is $\alpha - \beta + \delta - \gamma = 0$ and thus the bond valences and calculated lengths in Å (observed values in parentheses) are found:

Sr–O(1)	$v = 6/21$	$d = 2.58$ (2.65)
Sr–O(2)	$v = 4/21$	$d = 2.73$ (2.71)
Be(1)–O(1)	$v = 12/21$	$d = 1.59$ (1.60)
Be(1)–O(2)	$v = 10/21$	$d = 1.66$ (1.66)
Be(2)–O(2)	$v = 2/3$	$d = 1.53$ (1.53)

This example is used to introduce a compact way to present all the information necessary to calculate the bond valences in crystals. This is done by writing out the number per formula unit of each kind of atom before its symbol in the connectivity matrix and likewise preceeding the symbol for the bond valence by the number of bonds of that type (for the same unit). Thus

	O(1)	3 O(2)
Sr	3 α	6 β
2 Be(1)	2 γ	6 δ
Be(2)	·	3 ε

For compounds $A_aB_bX_xY_yZ_z$ there may be as many as six kinds of bond and four independent bond valence sums at the atoms. Accordingly two circuit equations are needed. Setting out the connectivity matrix (and establishing the notation for the individual valences):

$$\begin{array}{cccccc} & X & & Y & & Z \\ A & \alpha & — & \beta & — & \gamma \\ & | & & | & & | \\ B & \delta & — & \varepsilon & — & \xi \end{array}$$

It can be determined that these equations are:

$$\alpha - \beta + \varepsilon - \delta = \beta - \gamma + \varsigma - \varepsilon = 0$$

An example is the structure of β-Ga_2O_3 [25] = Ga(1)Ga(2)O(1)O(2)O(3). The connectivity matrix is

	O(1)	O(2)	O(3)
Ga(1)	α	2 β	γ
Ga(2)	2 δ	ε	3 ς

The calculated bond valences and distances (observed in parentheses) in Å are:

$\alpha = 62/75 \quad d = 1.80\ (1.80)$

$\beta = 56/75 \quad d = 1.84\ (1.83)$

$\gamma = 51/75 \quad d = 1.87\ (1.85)$

$\delta = 44/75 \quad d = 1.93\ (1.98)$

$\varepsilon = 38/75 \quad d = 1.98\ (1.94)$

$\varsigma = 33/75 \quad d = 2.03$ (mean 2.06)

Note that with the exception of δ and ε the bond lengths are correctly ordered in the calculation. Agreement any better than that observed is not really to be expected; in

α-Ga_2O_3 [26] there are only two kinds of atom but two Ga–O bond lengths of 1.92 and 2.08 Å – one would predict that they were equal using only bond valence considerations.

A final example must suffice here. A large number of "inverse spinels" are known with composition AB_2O_4. The coordinations are AO_6, BO_4 and BO_6 and each oxygen atom is coordinated to one tetrahedral and three octahedral metal atoms. There must be at least two kinds of O atom e.g. with coordinations $\{O\}^{iv}B^{vi}B_2A$ and $\{O\}^{iv}B^{vi}BA_2$ (other possibilities are $\{O\}^{iv}BA_3$ and $\{O\}^{iv}B^{vi}B_3$). The structure of Li_2TeO_4 (= $^{iv}Li^{vi}Li^{vi}TiO_4$) is known [27] and $LiZnNbO_4$ (= $^{vi}Li^{iv}Zn^{vi}NbO_4$) and Zn_2TiO_4 (= $^{iv}Zn^{vi}Zn^{vi}TiO_4$) are probably isostructural. The general formula for this structure type is $ABCX_2Y_2$ (C is the tetrahedrally-coordinated cation) with connectivity matrix

	$2X$	$2Y$
A	2α	4β
B	4γ	2δ
C	2ε	2ς

The methods outlined above give for the bond valences:

$$v_{AX} = \alpha = (7V_A - 8V_B - 6V_C + 24V_X)/66$$

$$v_{AY} = \beta = (13V_A + 4V_B + 3V_C - 12V_X)/66$$

$$v_{BX} = \gamma = (-2V_A + 7V_B - 3V_C + 12V_X)/66$$

$$v_{BY} = \delta = (4V_A + 19V_B + 6V_C - 24V_X)/66$$

$$v_{CX} = \varepsilon = (-V_A - 2V_B + 4V_C + 6V_X)/22$$

$$v_{CY} = \varsigma = (V_A + 2V_B + 7V_C - 6V_X)/22$$

Some special cases are considered later when the **inverse spinel** and **olivine** structures are discussed further.

7 Justification of the Ring and Circuit Sum Procedures and Further Examples

Earlier a general four-membered ring A–X–B–Y was considered and it was shown that after taking into account the independent valence sums at the atoms, the differences between the valences of the bonds formed by A and B could be expressed as:

$$\delta_A = v_{AX} - v_{AY} = 2s - V_A$$

$$\delta_B = v_{BX} - v_{BY} = 2V_X - V_B - 2s$$

so that

$$\delta_A + \delta_B = 2\,V_X - V_A - V_B = a \text{ (a constant)}$$

and therefore one could write

$$\delta_A = a - \varepsilon, \quad \delta_B = a + \varepsilon$$

The ring sum rule gives $\varepsilon = 0$.

To justify this it is supposed that, if subject to no other constraints, bonds from an atom strive to be as nearly equal in valence as possible (as in CH_4 and SF_6 etc.)[8] and further that the energy increases faster than linearly with difference in valence. If the energy varies as the nth power of the bond valence difference so that

$$\Delta E \sim (\delta_A)^n + (\delta_B)^n = (a - \varepsilon)^n + (a + \varepsilon)^n$$

if $n > 1$, ΔE is a minimum for $\varepsilon = 0$.

The procedure used by Brown [15] to determine bond valences in complex structures was to minimize the sum of the squares of the δ's, which is equivalent to setting $\varepsilon = 0$ (the ring sum rule).

The discussion of six- or higher-membered rings is not quite so straightforward but presumably a similar principle holds. In practice in most crystal structures one need only consider four-membered circuits on the connectivity diagram. The simplest structure examined so far [28] in which one has to consider six-membered rings is that of magnetoplumbite ($PbFe_{12}O_{19}$) in which there are eleven different types of atom.

It is emphasized again that sums around rings usually reduce to simpler sums around circuits. A good example is provided by the structure of the low-temperature form of Li_3PO_4 which is another structure that is a superstructure of **wurtzite.** The connectivity matrix is:

	O(1)	O(2)	2 O(3)
P	α	β	2γ
Li(1)	δ	ε	2ς
2 Li(2)	2η	2θ	4ι

and there are four independent circuits such as $\alpha\beta\varepsilon\delta$. The actual structure contains ten distinct six-membered rings, but it can readily be shown that the bond valence equations that they provide are all expressible in terms of the four basic circuit equations[9]. It is amusing that the high-temperature structure of Li_3PO_4 has an identical connectivity matrix so we would predict the same bond valences for it. The two structures are quite distinct topologically (the high temperature structure has four-membered Li–O–Li–O rings that are absent in the low-temperature form).

The Li_3PO_4 structure illustrates another general principle that often greatly simplifies a problem. It is this: If all the entries in a column (or row) are the same [e.g. the entries

8 Notable exceptions [compounds of d^n (n = 0, 4, 8, 9) and s^2 cations] are discussed later

9 It is not difficult, but tedious, to provide a fairly rigorous proof that all rings in a structure will provide equations that will be linear combinations of circuit equations

for O(1) and O(2)] or in a constant ratio [e.g. the entries for O(2) and O(3)] then the valences in each row (column) of the two columns (rows) are the same. This is demonstrated for the m and nth columns of a connectivity matrix in which a general entry is $a_{ij}\alpha_{ij}$ – here a_{ij} is the number of bonds of valence α_{ij}. Let there be b_m atoms of valence V_m in the mth column. The valence sums at atoms in the m and nth columns is

$$\Sigma_i a_{im}\alpha_{im} = b_m V_m$$

$$\Sigma_i a_{in}\alpha_{in} = b_n V_n$$

as (by hypothesis) $a_{im}/a_{in} = (b_m V_m)/(b_n V_n)$, these two equations combine to give:

$$\Sigma_i a_{im}(\alpha_{im} - \alpha_{in}) = 0$$

On the other hand there will be circuit equations of the sort:

$$\alpha_{im} - \alpha_{in} = \alpha_{i+1,m} - \alpha_{i+1,n}$$
(= constant for all i)

So that $\alpha_{im} = \alpha_{in}$.

In the case of Li_3PO_4 the connectivity matrix reduces to

	4 O
P	4 α
3 Li	12 β

and $\alpha = 5/4$, $\beta = 1/4$.

In the olivine structure there are six kinds of atom $ABCXYZ_2$ (Mg_2SiO_4 is a well known example). The connectivity matrix for the case of an oxide (in which A is at the centers of symmetry and C is four-coordinated) is:

Table 2. Bond valences in oxide olivine and inverse spinel structures. The numbers in the first three columns are the valences of the atoms A, B, C respectively (C is the tetrahedrally-coordinated cation). δ is the difference in valence for the M–O(1) and M–O(2) bonds

	A–O(1)	*A*–O(2)	*B*–O(1)	*B*–O(2)	*C*–O(1)	*C*–O(2)	δ
N N 8-2 *N*	*N*/6	*N*/6	*N*/6	*N*/6	2-*N*/2	2-*N*/2	0
1 2 5	9/66	12/66	21/66	24/66	27/22	28/22	1/22
1 3 4	7/66	13/66	31/66	37/66	21/22	23/22	2/22
1 4 3	5/66	14/66	41/66	50/66	15/22	18/22	3/22
1 5 2	3/66	15/66	51/66	63/66	9/22	13/22	4/22
1 6 1	1/66	16/66	61/66	76/66	3/22	8/22	5/22
2 3 3	20/66	23/66	32/66	35/66	16/22	17/22	1/22
2 4 2	18/66	24/66	42/66	48/66	10/22	12/22	2/22
2 5 1	16/66	25/66	52/66	61/66	4/22	7/22	3/22
3 4 1	31/66	34/66	43/66	46/66	5/22	6/22	1/22

	O(1)	O(2)	2 O(3)
A	α	β	$4\,\gamma$
B	$2\,\delta$	$2\,\varepsilon$	$2\,\varsigma$
C	η	θ	$2\,\iota$

from which it is immediately deduced that bonds from a given atom to O(1) and to O(2) have the same valence. The connectivity matrix is now written (redefining symbols for the bond valences) as:

	2[O(1), O(2)]	2 O(3)
A	$2\,\alpha$	$4\,\beta$
B	$4\,\gamma$	$2\,\delta$
C	$2\,\varepsilon$	$2\,\varsigma$

The solution for this case is readily obtained as a special case (with $V_X = V_Y = 2$) of the solution of the inverse spinel structure given above and is given in Table 2. [Note that, in the table, I write [O(1), O(2)] for olivine as O(1); and O(3) of olivine as O(2).] It may be verified that if further, $V_A = V_B = 2$ then $\alpha = \beta = \gamma = \delta = 1/3$ and $\varepsilon = \varsigma = 1$, so that in (e.g.) Mg_2SiO_4 all Mg–O and all Si–O bonds are found to have the same valence – one must therefore attribute the variations in bond lengths actually observed to factors other than bond valence constraints. The olivine structure is further discussed below.

8 Pauling's Electrostatic Valence Sum Rule

As defined by Pauling [10] the electrostatic strength of a bond from a cation to an anion is the cation charge (equal to its valence) divided by the number of bonds formed by the cation. In a coordination polyhedron with all bond lengths equal, the electrostatic bond strength, s, is equal to the bond valence, v. Pauling's rule states that the sum of the electrostatic bond strengths at an anion is "exactly or nearly equal" to the valence of the anion in a stable "ionic" structure. Writing s_{iX} for the Pauling bond strengths at a given anion and defining $\Delta p_X = V_X - \Sigma_{(i)} s_{iX}$, Pauling's rule is succintly stated as

$$\Delta p_X \sim 0$$

In compounds in which Δp_X is zero at every anion, the calculated bond valences are equal to the Pauling strength (olivine Mg_2SiO_4 is an example), i.e. $v = s$. When $\Delta p_X < 0$, $v > s$ for bonds to that anion and when $\Delta p_X > 0$, $v < s$. As bond lengths calculated from ionic radii are corrected for cation coordination, they are really bond lengths for a certain s and will only give correct values for $s = v$. This has been recognized particularly by Baur [29] who has described methods for correcting bond lengths calculated from ionic radii based on coordination numbers for the effect of Δp_X different from zero. It is worth noting that electrical neutrality requires that the sum of all the Δp_X for the anions in a structure is zero.

Pauling's rule is interpreted as follows. It states that in the most stable compounds $v = s$ – thus for a given cation bonded to anions of the same valence in such a compound, the bond valences are all equal. It has been seen that this can not be true in general for all structures owing to the constraints of the bond valence sums at the atoms. When v is not equal to s one uses the ring valence sums, a procedure that is equivalent to requiring that the differences in valences are as small as possible[10]; and a procedure that is compatible with Pauling's rule which implies that the more individual bond valences differ, the less stable is the compound.

There is an asymmetry to Pauling's rule in that it treats cations and anions differently. This is neither in the spirit of the bond valence method, nor satisfying when one considers the large number of structures and antistructures [3]. In some instances it leads to predictions that are probably wrong[11]. For this reason it is found useful to restate "Pauling's rule" in a different (and not entirely equivalent) way as follows:

The most stable structures are those in which the bonds, from a given atom to neighbors having the same valence as each other, have the same bond valence.

Thus consider $^{viii}La^{iv}O^{iv}F$ – the electrostatic strength of the La–O and La–F bonds is $s = 3/8$ so $\Delta p_F = 1/2$ and $\Delta p_O = -1/2$. However the La–O bond valences are all equal (1/2) as are the La–F bond valences (1/4) and such a compound is expected to be "stable" in contrast to the predictions of Pauling's rule.

Instead of using $\langle|\Delta p_X|\rangle$ as an index of "undesirability" of a structure I prefer to use the differences in bond valence. Using $^{iv}B_2^{ii}O^{iii}O_2$ as an example; bonds from the two-coordinated O atom have $v_{BO} = 1.0$ and from the three-coordinated atom $v_{BO} = 2/3$. The difference in the two B–O bond valences is thus 1/3. This is quite a large value for an oxide, and helps to explain why the B_2O_3 (II) structure – or any other structure with tetrahedral coordination[12] of the cations – is not found for compounds such as Al_2O_3, Ga_2O_3 etc. Note that the stable (zero pressure) structure of B_2O_3 has ^{iii}B and hence ^{ii}O. In this instance Pauling's rule leads to the same conclusion because $\langle|\Delta p_O|\rangle = 1/3$ for B_2O_3(II). In the isostructural $^{iv}Si_2^{iii}N_2^{ii}O$ all the bond valences are equal to unity.

9 The Olivine and "Inverse Spinel" Structures

We have alluded to the **olivine** and **inverse spinel** structures above. Here they are considered in the light of our discussion of Pauling's rule. As shown above, if all the anions are oxygen, the reduced connectivity matrix for the olivine structure and that for the inverse spinel structure are the same (the actual structures are of course quite different). The

10 We actually showed that the sum of the differences raised to some power greater than one must be a minimum

11 In the absence of a quantification of "stability" as used in this context, it is hard to make precise statements about the success or otherwise of Pauling's rule

12 If the cation is four coordinated, the only possible anion coordinations are one of two-coordination and two of three-coordination for this stoichiometry

correspondence is O(1) + O(2) for **olivine** → O(1) for **inverse spinel** and O(3) for **olivine** → O(2) for **inverse spinel.** There are six bond valences which are set out in Table 2. In the table A and B refer to octahedrally coordinated metal atoms and C is the tetrahedrally coordinated atom.

Also listed in Table 2 is δ, equal to the differences in bond valence for the bonds from metal atoms to O(1) and O(2). The larger the value of δ, the less stable the compound is expected to be on the basis of "Pauling's rule" (as restated above). One would expect that if all other factors were equal (i.e. no specific "preference" for octahedral or tetrahedral coordination), distributions with valences of ABC = 152 or 161 would be less stable than 125 or 116 – i.e. there is a site preference that depends on the valences of the other cations in the structure as well as an intinsic site preference of individual atoms. If the compound can be made, the larger the value of δ, the greater the distortion expected in the oxygen polyhedra surrounding the metal atoms. δ is in fact proportional to the difference in valence of the two octahedrally-coordinated metal atoms so that the compounds with large δ are also those most likely to be ordered.

The above statements are generally in accord with experience. **Olivines** and **spinels** $ABCO_4$ are known with valences of A and B (the octahedrally-coordinated atoms) the same and equal to 1, 2 or 3 (i.e. all possible values). When the valences of A and B are not equal olivines are apparently restricted to valences of ABC = 125, 134, 233. **Spinels** of the type 161 and 152 are known (Li_2TeO_4 and $LiZnNbO_4$) and are ordered. The occurrence of these compounds is presumably related to a strong "preference" of Te(VI) for octahedral and Zn for tetrahedral coordination.

10 Apparent Valences

A major interest in being able to calculate bond valences in crystals is to calculate bond lengths. One then uses bond valence – bond length correlations empirically determined from the available structure base. The reverse of this process is to use observed bond lengths to calculate apparent bond valences in crystal structures; these can then be summed at each atom to get the apparent atomic valence. If the correlation between length and valence were exact, the apparent valences so calculated would be the true valences. In practice there are usually small differences between the two quantities and, more rarely, large ones.

There are three well-established uses of the calculation of atomic valences. The first is simply as a check on the accuracy of a structure – which may be suspect if apparent valences are very different from the normal valence; some examples given below illustrate the typical agreement observed. The second is to locate atoms such as O and F which are difficult to distinguish by X-ray diffraction, or to distinguish between O, OH and H_2O when H atom positions are not known [30].

An example is provided by the structure of topaz $Al_2SiO_4F_2$ [31] the connectivity matrix is:

	O(1)	O(2)	2 O(3)	2 F
2 Al	2 α	2 β	4 γ	4 δ
Si	ε	ς	2 η	.

It can be seen that the valences in the first three columns of the matrix must be equal, i.e. that $\alpha = \beta = \gamma$ and $\varepsilon = \varsigma = \eta$. The valences are then equal to the Pauling strengths – all the bonds from Al have $v = 1/2$ and the bonds from Si have $v = 1$. The observed bond lengths and apparent valences (in parentheses) are:

anion *(X)*	Si–*X*	Al–*X*	$V' = \Sigma v'$
O(1)	1.636 (0.97)	1.896 (0.52)	2.00
O(2)	1.644 (0.95)	1.901 (0.51)	1.97
O(3)	1.642 (0.95)	1.884 (0.53) 1.897 (0.51)	2.00
F		1.790 (0.52) 1.800 (0.50)	1.02
$V' = \Sigma v'$	3.82	3.09	

In the sums the number of each kind of bond to each atom (obtained from the connectivity matrix) has been taken into account, and there are round-off errors. The agreement between V' and V is typical for a "good" case. This example is of historical interest [32] as the identification of F was made using Pauling's rule[13]. One could verify that the location of F was correct by interchanging F with O(1) or O(2) and repeating the calculation of V'_F; a result quite different from unity would be obtained.

A third use of calculated valences is to distinguish between different oxidation states of transition metal atoms. Often (not always) it is a useful approximation to assume a constant value of R_0 for a given atom in different oxidation states. Recent applications include: for Mo compounds, Bart and Ragaini [33], Ti(III)/Ti(IV) oxides [34, 35], Cu(II)/Cu(III) oxides [22] and W(V)/W(VI) oxides [36]. The last authors suggest that one adjust the parameter R_0 for each compound to make the total calculated valence equal to the correct value – thus in e.g. $W_{18}O_{49}$ the total valence of all the W atoms in a formula unit must add up to 98. One often finds non-integral valences by this procedure. The reason is that in electronically-conducting oxides the rate of electron transfer between atoms may exceed the rate at which atoms can relax to accommodate a change of valence, and one then has bond lengths that are intermediate between those appropriate for two different bond valences [37].

A simple example illustrates the method. Ilmenite, $FeTiO_3$ is a common mineral and the magnetic properties of its solid solutions with Fe_2O_3 are of considerable interest. There was at one time a lively debate [38] as to whether it was Fe(II)Ti(IV)O_3 or Fe(III)Ti(III)O_3 although the former valences are indicated from physical properties [39]. The bond lengths [40] are d(Ti–O) = 1.878 (3 ×) and 2.087 (3 ×) and d(Fe–O) = 2.070 (3 ×) and 2.197 Å (3 ×). From these we can evaluate the apparent valences of Ti and Fe (again using the Brown and Altermatt parameters) as set out below:

13 Good scientists are usually lucky, and Pauling was lucky that in topaz $s = v$ so his rule worked. It generally is not applicable to mixed-anion compounds as the example of LaOF showed

hypothesis	Fe(II), Ti(IV)	Fe(III), Ti(III)
V'_{Fe}	2.07	2.21
V'_{Ti}	3.97	3.65

The bond valence sums clearly indicate the valences Fe(II) and Ti(IV).

This work is more concerned with a fourth use of apparent atomic valences. The interest is in compounds in which there is no ambiguity concerning true valences and whose structures are (presumably) not in question. It is then asked why the apparent valence differs from the true valence on the occasions when it does so significantly. The hypothesis, clearly anticipated by Zachariasen [13] and, for molecules, by Bartell [41], is that bond lengths are influenced not only by valence but also by next-nearest ("non-bonded") interactions [42].

For a given type of bond (e.g. Ca–O), one can use known structures (for this example, the set of all good structures in which there are Ca–O bonds), to determine parameters for a bond length – bond valence equation. If bonds are affected by factors such as non-bonded interactions one will have some sort of "average equation" (biased perhaps by the fact that certain types of compound are more likely to be studied than others). It is supposed that bonds are generally lengthened by non-bonded repulsions. Then in a compound in which such repulsions are unusually large, one will find bonds lengthened more than average, and calculations from such bond lengths using the "average equation" will yield apparent valences less than the true valences. It follows that there must be another subset of structures with apparent valences greater than the true valence.

The specific examples discussed are oxides; the reason is that the large available data base has been critically examined by Brown and his collaborators and unbiased (except perhaps in the sense mentioned above) "average equations" have been published for bonds from many elements to oxygen [14].

The greatest range of apparent valence will come when a central atom is highly coordinated by a large number of "large" atoms and when the atoms in question have low valence (so that the individual bonds are weak and thus most susceptible to perturbations). Hence in oxides we expect to find the greatest effects for bonds to alkali and alkaline earth elements, and the smallest effects for bonds to atoms such as Si and P which have high valence and which form relatively few bonds.

Elsewhere [43] it has been argued that in metal-rich compounds, metal ... metal non-bonded interactions reduce the stability of the compound as measured by its heat of atomization. It might be expected that these effects also show up in comparing the apparent valences V' with the true valences V. It was in fact shown earlier [42] that in the alkali metal oxides M_2O, the bond lengths are significantly longer than would be expected from bond length – bond valence correlations derived for alkali metal oxides where the metal/oxygen ratio was less than one. Thus the ratio V'_M/V_M would be found to be less than unity. The ratio $\Gamma = V'/V$ has been calculated for a number of simple oxides and listed in Table 3.

The results are rather revealing. They show that in general Γ is significantly less than unity for the alkali and alkaline earth elements and correlates generally with the trends in bond energies of the binary compounds as compared with bond energies in ternary etc.

Table 3. Ratio of apparent valence to true valence ($\Gamma = V'/V$) for cations in some binary oxides

Oxide	Γ	Oxide	Γ	Oxide	Γ	Oxide	Γ
Li_2O	0.95	BeO	0.97	B_2O_3	1.02		
Na_2O	0.79	MgO	0.98	Al_2O_3	0.99	SiO_2	1.04
K_2O	0.67	CaO	0.92	Sc_2O_3	0.98	GeO_2	1.04
Rb_2O	0.67	SrO	0.88	Y_2O_3	0.98	SnO_2	1.00
Cs_2O	0.91	BaO	0.81	La_2O_3	0.96[a]	PbO_2	1.05

[a] see also the discussion of La_2O_3 in the text

oxides [43, 44]. The periodic trend is also clearly established with Γ being smallest at the bottom left of the periodic table. Cs_2O is apparently exceptional, but note (i) that the coordination of O by Cs is lower (six) than in the other alkali oxides (where it is eight) and (ii) that it is difficult to obtain very good bond length – bond valence correlations for Cs oxides [42]. Thus the trend appears to be that when oxygen is highly coordinated by "large" metal atoms Γ is significantly less than unity; on the other hand when oxygen is in low coordination (e.g. 2-fold in SiO_2), Γ is close to or exceeds unity. These observations are entirely in accord with ideas [43] concerning the role of metal . . . metal interactions in influencing the structures and stabilities of oxides.

The case of La_2O_3 is particularly informative. In the structure there are two kinds of O atom with coordinations OLa_4 and OLa_6 [cf. Ref. 3]. Using the Brown and Altermatt value of $R_0 = 2.172$ Å one calculates the bond lengths to be $d(^{iv}O\text{–}La) = 2.43$ Å and $d(^{vi}O\text{–}La) = 2.58$ Å. The lengths observed for the bonds to the tetrahedral O are fairly close to that calculated, so that $V'(^{iv}O) = 2.22$. However, the bonds to the octahedral oxygen are found to be much longer (2.73 Å) than calculated[14], and $V'(^{vi}O) = 1.33$. If the O–La bonds had the calculated lengths the La . . . La distances around the tetrahedral O would be 3.97 Å (observed 3.94 and 3.82 Å) and the La . . . La distances around the octahedral oxygen would be 3.65 Å (observed again 3.94 and 3.82 Å). These distances might be compared with the nearest-neighbor distances of 3.77 and 3.74 Å in La metal which has a eutactic ("close packed") arrangement just as in the oxide.

A comparison with the bond lengths in $La(OH)_3$ [48] is also revealing. Here the coordinations are $\{La\}O_9$ and $\{O\}La_3H$. The bond lengths from La are 2.59 (6 ×) and 2.55 (3 ×) Å and $V'_{La} = 3.03$. The bonds have the same formal valence (1/3) as those from the octahedral O in La_2O_3 yet they are about 0.16 Å shorter!

The above observations of "normal" bond lengths to tetrahedral O are consistent with the observation [49] that in many oxy-salts of La the tetrahedral La_2O_2 layer persists as a structural unit. On the other hand the low bond valence sum at the octahedral oxygen is consistent with the known properties of La_2O_3 which, like the alkali metal and heavier alkaline earth oxides, rapidly reacts with atmospheric water or CO_2 at room temperature.

One is left with the question, concerning La_2O_3, of why the *A* type structure is formed at all when the alternative *C* type structure (with all four-coordinated O) is available. The answer is beyond the scope of the present paper, but it is planned to at least address the question elsewhere (O'Keeffe and Hyde, to be published).

14 It is for this reason that Zachariasen [45] suspected that the La_2O_3 structure might be incorrect. However, the structure of it [46] and the isostructural Ce_2O_3 [47] have been carefully refined

A study has been made of bond lengths in Ca oxy-compounds for which there is a large number of good structures. The bond valence – bond length parameters derived were very close to those of Brown and Altermatt so I continue to use their values. Table 4 lists $\Gamma = V'/V$ for those compounds in which there are either just one kind of Ca and O atom or those such as γ-Ca_2SiO_4 in which there are more than one kind of Ca but with the same coordination. For the latter the average value of Γ is listed although considerable interest attaches to the differences in apparent valence of the different atoms in compounds such as Ca_2SiO_4 [50]. It should be noted that in all the compounds other than CaO, O is coordinated by four or fewer atoms.

Several interesting points emerge from these results. It may be noted that in compounds with a given anion coordination the Ca–O bonds get longer (Γ_{Ca} gets smaller) as a more electronegative atom is replaced with a more electropositive one. Thus compare the **garnets** with {O}Ca_2AlSi and {O}Ca_2ScSi or the pair with {O}Ca_2ScSi and {O}Ca_2ScGe. Likewise in the compounds $CaCu_3M_4O_{12}$, compare the values of Γ_{Ca} in the compounds with {O}$CaCuGe_2$ and {O}$CaCuTi_2$. A plausible interpretation of these observations (and one in accord with previous ideas developed by O'Keeffe and Hyde [44]) is that the weaker Ca–O bonds are extended by increasing repulsions as larger cations are included in the coordination around O.

A second point of interest that emerges from Table 4 is that in compounds with several cations but only one kind of anion, there is a compensating effect that makes Γ_O close to unity. This is illustrated in more detail for the **garnets** $A_3Al_2Si_3O_{12}$:

Table 4. Ratio of apparent to true valence in some calcium compounds

Compound	Ca coord.	Γ_{Ca}	O coord.	Γ_O
CaO	{Ca}O_6	0.92	{O}Ca_6	0.92
$Ca(OH)_2$	{Ca}O_6	1.03	{O}Ca_3H	1.02
γ-Ca_2SiO_4	{Ca}O_6	1.03[a]	{O}Ca_3Si	1.00[a]
$CaCO_3$ (calcite)	{Ca}O_6	1.09	{O}Ca_2C	1.02
$CaMg(CO_3)_2$	{Ca}O_6	0.99	{O}CaMgC	1.01
$Ca_3Al_2Si_3O_{12}$	{Ca}O_8	1.26	{O}Ca_2AlSi	1.03
$Ca_3Sc_2Si_3O_{12}$	{Ca}O_8	1.08	{O}Ca_2ScSi	1.00
$Ca_3Fe_2Ge_3O_{12}$	{Ca}O_8	1.20	{O}Ca_2FeGe	0.99
$Ca_3Sc_2Ge_3O_{12}$	{Ca}O_8	1.02	{O}Ca_2ScGe	0.99
$CaWO_4$	{Ca}O_8	1.06	{O}Ca_2W	0.99
$CaCO_3$ (aragonite)	{Ca}O_9	1.01	{O}Ca_3C	1.00[a]
$CaCu_3Ge_4O_{12}$	{Ca}O_{12}	1.25	{O}$CaCuGe_2$	1.01
$CaCu_3Mn_4O_{12}$	{Ca}O_{12}	1.20	{O}$CaCuMn_2$	0.99
$CaCu_3Ti_4O_{12}$	{Ca}O_{12}	1.07	{O}$CaCuTi_2$	1.00
$CaGeO_3$	[b]	1.19		
$CaTiO_3$	[b]	1.04		
$CaSnO_3$	[b]	0.98		
$CaZrO_3$	[b]	0.93		

[a] average of two or more different atoms with the same coordination
[b] irregular – first twelve oxygen neighbors counted

A	Γ_A	Γ_{Al}	Γ_{Si}	Γ_O
Ca	1.26	0.96	0.95	1.03
Mg	0.86	1.06	0.97	0.97

Likewise when there is only one kind of cation, but very different anions, the compensation effect still appears as may be illustrated by the example of La_2O_3 already adduced:

$\Gamma_{^{iv}O} = 1.11$ $\Gamma_{^{vi}O} = 0.66$ $\Gamma_{La} = 0.96$

Effects of this nature will have to be taken into account in approaches to the prediction of bond lengths in crystals when there are easily-perturbable bonds to large electropositive atoms. The thesis here is of course that deviations of Γ from unity in such compounds are diagnostic of short-range repulsions in the crystal.

11 Limitations to Predictions of Bond Lengths

Although the bond valence method is clearly superior to older methods using sums of radii for predicting bond lengths, it is far from perfect. As discussed above, in predicting bond lengths to (especially) the more electropositive elements in oxides, some method of accounting for atom . . . atom non-bonded repulsions and other factors that perturb bond lengths will have to be devised.

More importantly, there are a number of cases where distorted polyhedra are formed around cations for electronic reasons. Chief among these are (a) transition metal atoms and (b) "lone pair" atoms. Each of these cases is briefly discussed.

Transition metal ions with a d^0 configuration [notably Ti(IV), Zr(IV), V(V), Nb(V), Ta(V), Mo(VI) and W(VI)] are often found in irregular coordination for reasons that are not apparent when considered from the point of view of the bond-valence method. WO_3 is a good example: in its several polymorphs W is bonded to six oxygen atoms and O to two W atoms, but the W–O bonds are of very unequal lengths, whereas the simple bond valence method (or sums of radii) would predict these to be all of the same length. In such a case a second order Jahn-Teller distortion allowing mixing of the highest occupied orbitals and the empty d manifold so that the former are lowered in energy [51][15]. The many ferroelectric perovskites containing these ions are manifestations of the same phenomenon.

In compounds of such elements one also finds gross violations of Pauling's electrostatic valence sum rule (and the restated rule given above). Notable examples are in the "shear" structures such as TiO_{2-x} and WO_{3-x} in which one finds all octahedrally-coordinated cations but O bonded to three and four Ti atoms or to two and to three W respectively. In MoO_3, again with octahedrally coordinated Mo, one finds O bonded to one, two and three Mo atoms with correspondingly a wide variation of bond lengths and with Pauling bond strength sums at O of one, two and three respectively. It is the special

15 This case might be contrasted with the topologically identical $ReO_3(d^1)$ and $TeO_3(d^{10})$ structures which have regular octahedra

property of d^0 cations that favors irregular coordination figures that allows such structures to be formed.

As is well known, other specific d electron configurations such as d^4, d^9 and low-spin d^8 also lead to distorted environments. This is often explained as a first order Jahn-Teller effect in the first two cases, but the distortion to lond and short bonds often occurs without change in symmetry as in $Ba_2YCu_3O_7$ [23]. Again, to predict bond lengths in such compounds, these effects must be explicitly considered. Other transition metal atoms, such as Fe(III), are however, quite tractable.

The remarks in the above paragraph also apply to "lone pair" atoms such as Sn(II), As(III), Te(IV), Tl(I), Pb(II) and Bi(III) which almost always are found in irregular coordination (e.g. three long and three short bonds). Presumably in these instances there is also a second-order Jahn-Teller distortion.

12 "Unusual" Valences

Even in some simple binary compounds, the valences deduced from the composition may be wrong. A simple example is afforded by CeN (with the **NaCl** structure) in which the bond length is much to short for Ce(III) bonded to six N atoms [52]. In fact the Ce–N bond length is the same as in Li_2CeN_2 (which has the **anti La_2O_3** structure [53] in which Ce(IV) is bonded to six N atoms. Accordingly CeN might be better written Ce(IV)N.e.

A second example is afforded by GdO_2 [54] which has the **PbFCl** structure. Discarding the unlikely hypothesis of Gd(IV) one finds for the apparent valences of O(1) $V' = 2.08$ and of O(2) $V' = 0.42$. The most plausible interpretation of these results is that O(2) is O^- and that one needs a different value of R for bonds from Gd(III) to O^- ($R = 2.39$) and to O^{2-} ($R = 2.07$).

13 Concluding Remarks

In this paper, I have outlined how the bond valence method may be used to predict bond lengths in crystal structures, particularly in cases where use of radius sums fails. I have also indicated some of its limitations. My feeling is that to go much beyond the present level of treatment would require the introduction of too many empirical parameters to be generally useful. On the other hand the inverse procedure of using observed bond lengths to calculate apparent valences promises to be very fruitful and capable of considerable further development.

The reader will have noticed that there are no structural drawings. This is because the bond valence method, as described here, relies solely on nearest neighbor connectivities, which are most succinctly set out in connectivity matrices. Many examples of the latter have been given.

As mentioned in the Introduction, the examples used to illustrate the method have been drawn mostly from the extensive literature on oxides. There is no reason to suppose that the method would not be equally applicable to a much wider class of materials.

However, to avoid circular reasoning, before making an application to a given structure, one needs an independent set of parameters for each type of bond in the structure. This in turn requires examination of a large set of good structures containing these bonds, and for which the atomic valences are unambiguous.

14 Appendix 1. The Bond Valence Method and Kirchhoff's Laws

We have mentioned that the valence sums at atoms and around rings in structures resemble the application of Kirchhoff's laws to the calculation of currents in electrical networks. Kirchhoff's Laws [55] relate the currents I flowing through a link of a network, the resistances, R, of those links and the applied e.m.f.'s E. There is the *junction law* that states that the algebraic sum of all the currents meeting at a junction is zero i.e.

$$\Sigma_{(\mathrm{junction})} I_i = 0 \tag{A1}$$

and the *loop law* that states that the sum of the RI terms around any closed path (a loop) is equal to the algebraic sum of the applied e.m.f.'s in the given path:

$$\Sigma_{(\mathrm{loops})} R_i I_i = \Sigma E_i \tag{A2}$$

On the other hand the bond valence sum rules are for rings

$$\Sigma_{(\mathrm{rings})} v_{ij} = 0 \tag{A3}$$

and at the atoms

$$\Sigma_{(\mathrm{atom}\ i)} n_{ij} v_{ij} = V_j \tag{A4}$$

where n_{ij} is the number of bonds of the ij type.

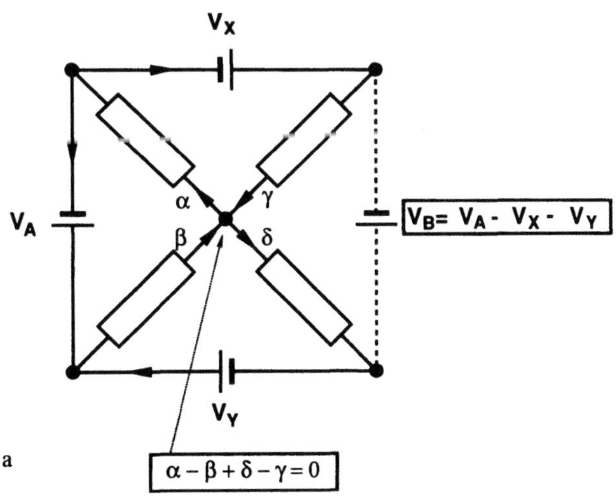

Fig. 1. Electrical circuit analog of a four-membered ring *A-X-B-Y*

Thus there is a formal correspondence if one makes the exchanges:

$$I \leftrightarrow v; \quad R \leftrightarrow n \quad \text{and} \quad \Sigma E \leftrightarrow V \tag{A5}$$

but this also requires

$$\text{junctions} \leftrightarrow \text{rings} \quad \text{and} \quad \text{loops} \leftrightarrow \text{atoms} \tag{A6}$$

Let us reconsider the A–X–B–Y ring for which the valence sums were given before (Sect. 4). Figure 1 shows an electrical network that reproduces the valence sums when currents are interpreted as bond valences and e.m.f.'s as atom valences and resistances are taken as unity. It should be clear that the correspondence between the current problem and the bond valence problem is only formal.

A second example shows the correspondence between a bond valence problem expressed in terms of the connectivity matrix and an electrical circuit. In the matrix there are x atoms of type X per formula unit etc. and the number of bonds with valence i is expressed as R_i:

	xX	yY
aA	$R_\alpha \alpha$	$R_\beta \beta$
bB	$R_\gamma \gamma$	$R_\delta \delta$
cC	$R_\varepsilon \varepsilon$	$R_\varsigma \varsigma$

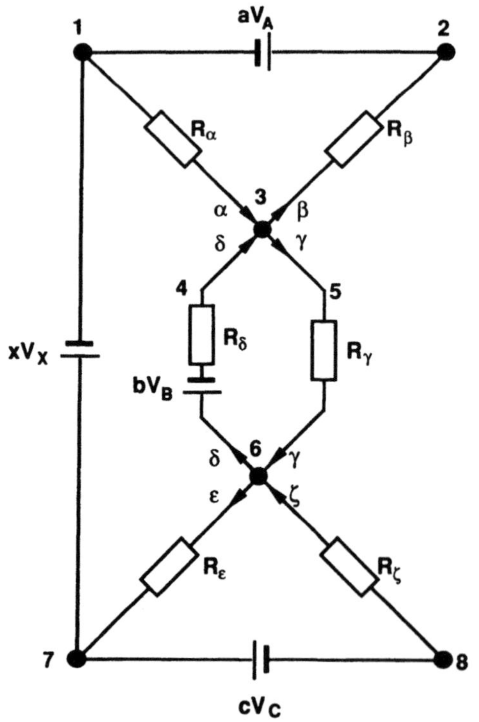

Fig. 2. An electrical circuit corresponding to the connectivity matrix discussed in Appendix 1

A circuit that yields the same equations for the bond valences is shown in Fig. 2. The equations derived from the figure and the connectivity matrix are:

loop	123	≡ sum at atom A	$R_\alpha\alpha + R_\beta\beta = aV_A$
loop	3465	≡ sum at atom B	$R_\gamma\gamma + R_\delta\delta = bV_B$
loop	678	≡ sum at atom C	$R_\varepsilon\varepsilon + R_\varsigma\varsigma = cV_C$
loop	13 567	≡ sum at atom X	$R_\alpha\alpha + R_\gamma\gamma + R_\varepsilon\varepsilon = xV_X$
junction	3	≡ circuit $\alpha\beta\delta\gamma$	$\alpha - \beta + \delta - \gamma = 0$
junction	4	≡ circuit $\gamma\delta\varsigma\varepsilon$	$\gamma - \delta + \varsigma - \varepsilon = 0$

It rapidly becomes difficult to derive electrical circuits for more complex connectivity matrices – perhaps impossible in some instances. For this reason I think that the analogy is probably not worth pursuing much beyond the insight that it gives in the case of the simple four-membered ring structure.

15 Appendix 2. Bond Valence Parameters

Useful list of bond valence parameters can be found in the compilations by Brown and Altermatt [14] and by Brown [2]. I have avoided any discussion of the physical significance of the parameters, and their relationship to other measures of "size" such as atomic and ionic radii. For a discussion of systematic variations of R across the periodic table reference should be made to the paper of Brown and Altermatt [14]. Interesting observations on this topic have also been made by Gibbs and Boisen [56] and by Burdett [57]. The development of parameters for many types of bond is a task still to be accomplished.

It is interesting that the values of R, which represent the length of a bond of unit strength are not in accord with traditional concepts of atom sizes. Thus considering bonds to oxygen, R for Na(I), Mg(II), Al(III), Si(IV) and P(V) are all almost the same (cf. [58]). For bonds to a given atom in different oxidation state R often *increases* with oxidation state (e.g. for Cu(I), Cu(II) and Cu(III) R = 1.59, 1.68 and 1.74 Å [23]) although the ionic radius decreases.

Acknowledgements. Many of the ideas presented here, and the specific examples adduced have developed over a number of years of extensive discussions with Bruce Hyde whose input (and restraint) have been essential. Staffan Hansen and Tim Wagner have also patiently helped me elaborate many of the arguments. This work is part of a program supported by the National Science Foundation (grant DMR 88 13524).

16 References

1 Brown ID (1978) Chem. Soc. Rev. 7: 359
2 Brown ID (1981) In: O'Keeffe M, Navrotsky A (eds) Structure and bonding in crystals II. Academic, New York
3 O'Keeffe M, Hyde BG (1985) Structure and Bonding 61: 79
4 Pauling L (1947) J. Amer. Chem. Soc. 69: 542

5 Byström A, Wilhelmi K-A (1951) Acta Chem. Scand. 5: 1003
6 Bragg WL (1920) Phil. Mag. 40: 169
7 Slater JC (1965) Quantum Theory of Molecules and Solids 2, McGraw-Hill, New York
8 Shannon RD, Prewitt CT (1969) Acta Crystallogr. B25: 925
9 O'Keeffe M (1981) In: O'Keeffe M, Navrotsky A (eds) Structure and bonding in crystals I. Academic, New York
10 Pauling L (1960) The nature of the chemical bond, Cornell University Press, Ithaca
11 Adamson A (1979) A textbook of physical chemistry, Academic, New York
12 Johnston HS (1966) Gas phase reaction theory, Ronald, New York
13 Zachariasen WH (1963) Acta Crystallogr. 16: 385
14 Brown ID, Altermatt D (1985) Acta Crystallogr. B41: 244
15 Brown ID (1977) Acta Crystallogr. B33: 1305
16 Mackay AL, Finney JL (1973) J. Appl. Crystallogr. 6: 284
17 Horiuchi H, Sawamoto H (1981) Amer. Mineral. 66: 568
18 O'Keeffe M, Hyde BG (1978) Acta Crystallogr. B34: 3519
19 Laurent Y, Guyader J, Roult G (1981) Acta Crystallogr. B37: 911
20 Marezio M (1965) Acta Crystallogr. B37: 481
21 Hesse K-F (1977) Acta Crystallogr. B33: 901
22 Völlenkle H, Wittmann J (1968) Mh. Chem. 99: 244
23 O'Keeffe M, Hansen S (1988) J. Am. Chem. Soc. 110: 1506
24 Harris LA, Yakel HL (1969) Acta Crystallogr. B25: 1647
25 Geller S (1960) J. Chem. Phys. 33: 676
26 Marezio M, Remeika JP (1967) J. Chem. Phys. 46: 1862
27 Daniel F, Moret J, Phillippot E, Maurin M (1977) J. Solid State Chem. 22: 113
28 Wagner TR, O'Keeffe M (1988) J. Solid State Chem. 73: 211
29 Baur W (1981) In: O'Keeffe M, Navrotsky A (eds) Structure and bonding in crystals II. Academic, New York
30 Donnay G, Allmann R (1970) Amer. Mineral. 55: 1003
31 Ribbe PH, Gibbs GV (1971) Amer. Mineral. 56: 24
32 Pauling L (1928) Proc. Nat. Acad. Sci. U.S.A. 14: 603
33 Bart JCJ, Ragaini V (1979) Inorg. Chim. Acta 36: 261
34 Le Page Y, Strobel P (1982) J. Solid State Chem. 47: 6
35 Trömel M (1983) Acta Crystallogr. B39: 664
36 Domengès B, McGuire NK, O'Keeffe M (1985) J. Solid State Chem. 56: 94
37 Robin MB, Day P (1967) Adv. Inorg. Chem. Radiochem. 10: 247
38 Hámos LV, Stscherbina W (1931) Nacr. Ges. Wiss. Gotting. Math. Phys. Kl. Fachgruppen 232
39 Ruby SL, Shirane G (1961) Phys. Rev. 123: 1239
40 Morosin B, Baughman RJ, Ginley DS, Butler MA (1978) J. Appl. Cryst. 11: 121
41 Bartell LS (1962) Tetrahedron 17: 177
42 McGuire NK, O'Keeffe M (1984) J. Solid State Chem. 54: 49
43 O'Keeffe M, Hyde BG (1984) Nature 309: 411
44 O'Keeffe M, Stuart JA (1983) Inorg. Chem. 22: 177
45 Zachariasen WH (1978) J. Less-Common Mets. 62: 1
46 Adelbert P, Traverse JP (1979) Mat. Res. Bull. 14: 303
47 Bärnighausen H, Schulter G (1985) J. Less-Common Mets. 110: 385
48 Beall GW, Milligan WO, Wolcott HA (1977) J. Inorg. Nucl. Chem. 39: 65
49 Caro P (1972) In: Roth RS, Schneider SJ (eds) Solid state chemistry, N.B.S. Spec. Publ. 364, Washington, p. 367
50 Barbier J, Hyde BG (1985) Acta Crystallogr. B41: 383
51 Burdett JK (1980) Molecular shapes, Wiley, New York
52 O'Keeffe M (1979) Acta Crystallogr. A35: 776
53 Halot D, Flahaut J (1971) C. R. Acad. Sci. Paris C272: 465
54 Imanov IM, Ragimli NA, Semiletov SA (1975) Soviet Phys. Crystallogr. 19: 466 [trans. of Krystallografiya 19: 751 (1974)]
55 Page L, Adams NI (1931) Principles of electricity, Van Nostrand, New York
56 Gibbs GV, Boisen MB (1987) Mat. Res. Soc. Symp. Proc. 73: 515
57 Burdett JK (1988) Chem. Revs. 88: 3
58 Brown ID, Wu KK (1976) Acta Crystallogr. B32: 1957

Author Index Volumes 1–71

Ahrland, S.: Factors Contributing to (b)-behaviour in Acceptors. Vol. 1, pp. 207–220.

Ahrland, S.: Thermodynamics of Complex Formation between Hard and Soft Acceptors and Donors. Vol. 5, pp. 118–149.

Ahrland, S.: Thermodynamics of the Stepwise Formation of Metal-Ion Complexes in Aqueous Solution. Vol. 15, pp. 167–188.

Allen, G. C., Warren, K. D.: The Electronic Spectra of the Hexafluoro Complexes of the First Transition Series. Vol. 9, pp. 49–138.

Allen, G. C., Warren, K. D.: The Electronic Spectra of the Hexafluoro Complexes of the Second and Third Transition Series. Vol. 19, pp. 105–165.

Alonso, J. A., Balbás, L. C.: Simple Density Functional Theory of the Electronegativity and Other Related Properties of Atoms and Ions. Vol. 66, pp. 41–78.

Ardon, M., Bino, A.: A New Aspect of Hydrolysis of Metal Ions: The Hydrogen-Oxide Bridging Ligand $(H_3O_2^-)$. Vol. 65, pp. 1–28.

Augustynski, J.: Aspects of Photo-Electrochemical and Surface Behaviour of Titanium(IV) Oxide. Vol. 69, pp. 1–61.

Averill, B. A.: Fe–S and Mo–Fe–S Clusters as Models for the Active Site of Nitrogenase. Vol. 53, pp. 57–101.

Babel, D.: Structural Chemistry of Octahedral Fluorocomplexes of the Transition Elements. Vol. 3, pp. 1–87.

Bacci, M.: The Role of Vibronic Coupling in the Interpretation of Spectroscopic and Structural Properties of Biomolecules. Vol. 55, pp. 67–99.

Baker, E. C., Halstead, G.W., Raymond, K. N.: The Structure and Bonding of 4*f* and 5*f* Series Organometallic Compounds. Vol. 25, pp. 21–66.

Balsenc, L. R.: Sulfur Interaction with Surfaces and Interfaces Studied by Auger Electron Spectrometry. Vol. 39, pp. 83–114.

Banci, L., Bencini, A., Benelli, C., Gatteschi, D., Zanchini, C.: Spectral-Structural Correlations in High-Spin Cobalt(II) Complexes. Vol. 52, pp. 37–86.

Bartolotti, L. J.: Absolute Electronegativities as Determined from Kohn-Sham Theory. Vol. 66, pp. 27–40.

Baughan, E. C.: Structural Radii, Electron-cloud Radii, Ionic Radii and Solvation. Vol. 15, pp. 53–71.

Bayer, E., Schretzmann, P.: Reversible Oxygenierung von Metallkomplexen. Vol. 2, pp. 181–250.

Bearden, A. J., Dunham, W. R.: Iron Electronic Configurations in Proteins: Studies by Mössbauer Spectroscopy. Vol. 8, pp. 1–52.

Bergmann, D., Hinze, J.: Electronegativity and Charge Distribution. Vol. 66, pp. 145–190.

Berners-Price, S. J., Sadler, P. J.: Phosphines and Metal Phosphine Complexes: Relationship of Chemistry to Anticancer and Other Biological Activity. Vol. 70, pp. 27–102.

Bertini, I., Luchinat, C., Scozzafava, A.: Carbonic Anhydrase: An Insight into the Zinc Binding Site and into the Active Cavity Through Metal Substitution. Vol. 48, pp. 45–91.

Blasse, G.: The Influence of Charge-Transfer and Rydberg States on the Luminescence Properties of Lanthanides and Actinides. Vol. 26, pp. 43–79.

Blasse, G.: The Luminescence of Closed-Shell Transition Metal-Complexes. New Developments. Vol. 42, pp. 1–41.

Blauer, G.: Optical Activity of Conjugated Proteins. Vol. 18, pp. 69–129.

Bleijenberg, K. C.: Luminescence Properties of Uranate Centres in Solids. Vol. 42, pp. 97–128.

Boča, R., Breza, M., Pelikán, P.: Vibronic Interactions in the Stereochemistry of Metal Complexes. Vol. 71, pp. 57–97.

Boeyens, J. C. A.: Molecular Mechanics and the Structure Hypothesis. Vol. 63, pp. 65–101.

Bonnelle, C.: Band and Localized States in Metallic Thorium, Uranium and Plutonium, and in Some Compounds, Studied by X-Ray Spectroscopy. Vol. 31, pp. 23–48.

Bradshaw, A. M., Cederbaum, L. S., Domcke, W.: Ultraviolet Photoelectron Spectroscopy of Gases Adsorbed on Metal Surfaces. Vol. 24, pp. 133–170.

Braterman, P. S.: Spectra and Bonding in Metal Carbonyls. Part A: Bonding. Vol. 10, pp. 57–86.

Braterman, P. S.: Spectra and Bonding in Metal Carbonyls. Part B: Spectra and Their Interpretation. Vol. 26, pp. 1–42.

Bray, R. C., Swann, J. C.: Molybdenum-Containing Enzymes. Vol. 11, pp. 107–144.

Brooks, M. S. S.: The Theory of 5f Bonding in Actinide Solids. Vol. 59/60, pp. 263–293.

van Bronswyk, W.: The Application of Nuclear Quadrupole Resonance Spectroscopy to the Study of Transition Metal Compounds. Vol. 7, pp. 87–113.

Buchanan, B. B.: The Chemistry and Function of Ferredoxin. Vol. 1, pp. 109–148.

Buchler, J. W., Kokisch, W., Smith, P. D.: *Cis, Trans,* and *Metal* Effects in Transition Metal Porphyrins. Vol. 34, pp. 79–134.

Bulman, R. A.: Chemistry of Plutonium and the Transuranics in the Biosphere. Vol. 34, pp. 39–77.

Bulman, R. A.: The Chemistry of Chelating Agents in Medical Sciences. Vol. 67, pp. 91–141.

Burdett, J. K.: The Shapes of Main-Group Molecules; A Simple Semi-Quantitative Molecular Orbital Approach. Vol. 31, pp. 67–105.

Burdett, J. K.: Some Structural Problems Examined Using the Method of Moments. Vol. 65, pp. 29–90.

Campagna, M., Wertheim, G. K., Bucher, E.: Spectroscopy of Homogeneous Mixed Valence Rare Earth Compounds. Vol. 30, pp. 99–140.

Ceulemans, A., Vanquickenborne, L. G.: The Epikernel Principle. Vol. 71, pp. 125–159.

Chasteen, N. D.: The Biochemistry of Vanadium, Vol. 53, pp. 103–136.

Cheh, A. M., Neilands, J. P.: The δ-Aminolevulinate Dehydratases: Molecular and Environmental Properties. Vol. 29, pp. 123–169.

Ciampolini, M.: Spectra of 3 d Five-Coordinate Complexes. Vol. 6, pp. 52–93.

Chimiak, A., Neilands, J. B.: Lysine Analogues of Siderophores. Vol. 58, pp. 89–96.

Clack, D. W., Warren, K. D.: Metal-Ligand Bonding in 3d Sandwich Complexes, Vol. 39, pp. 1–41.

Clark, R. J. H., Stewart, B.: The Resonance Raman Effect. Review of the Theory and of Applications in Inorganic Chemistry. Vol. 36, pp. 1–80.

Clarke, M. J., Fackler, P. H.: The Chemistry of Technetium: Toward Improved Diagnostic Agents. Vol. 50, pp. 57–78.

Cohen, I. A.: Metal-Metal Interactions in Metalloporphyrins, Metalloproteins and Metalloenzymes. Vol. 40, pp. 1–37.

Connett, P. H., Wetterhahn, K. E.: Metabolism of the Carcinogen Chromate by Cellular Constitutents. Vol. 54, pp. 93–124.

Cook, D. B.: The Approximate Calculation of Molecular Electronic Structures as a Theory of Valence. Vol. 35, pp. 37–86.

Cotton, F. A., Walton, R. A.: Metal-Metal Multiple Bonds in Dinuclear Clusters. Vol. 62, pp. 1–49.

Cox, P. A.: Fractional Parentage Methods for Ionisation of Open Shells of *d* and *f* Electrons. Vol. 24, pp. 59–81.

Crichton, R. R.: Ferritin. Vol. 17, pp. 67–134.

Daul, C., Schläpfer, C. W., von Zelewsky, A.: The Electronic Structure of Cobalt(II) Complexes with Schiff Bases and Related Ligands. Vol. 36, pp. 129–171.

Dehnicke, K., Shihada, A.-F.: Structural and Bonding Aspects in Phosphorus Chemistry-Inorganic Derivates of Oxohalogeno Phosphoric Acids. Vol. 28, pp. 51–82.

Dobiáš, B.: Surfactant Adsorption on Minerals Related to Flotation. Vol. 56, pp. 91–147.

Doi, K., Antanaitis, B. C., Aisen, P.: The Binuclear Iron Centers of Uteroferrin and the Purple Acid Phosphatases . Vol. 70, pp. 1–26.

Doughty, M. J., Diehn, B.: Flavins as Photoreceptor Pigments for Behavioral Responses. Vol. 41, pp. 45–70.

Drago, R. S.: Quantitative Evaluation and Prediction of Donor-Acceptor Interactions. Vol. 15, pp. 73–139.

Duffy, J. A.: Optical Electronegativity and Nephelauxetic Effect in Oxide Systems. Vol. 32, pp. 147–166.

Dunn, M. F.: Mechanisms of Zinc Ion Catalysis in Small Molecules and Enzymes. Vol. 23, pp. 61–122.

Emsley, E.: The Composition, Structure and Hydrogen Bonding of the β-Deketones. Vol. 57, pp. 147–191.

Englman, R.: Vibrations in Interaction with Impurities. Vol. 43, pp. 113–158.

Epstein, I. R., Kustin, K.: Design of Inorganic Chemical Oscillators. Vol. 56, pp. 1–33.

Ermer, O.: Calculations of Molecular Properties Using Force Fields. Applications in Organic Chemistry. Vol. 27, pp. 161–211.

Ernst, R. D.: Structure and Bonding in Metal-Pentadienyl and Related Compounds. Vol. 57, pp. 1–53.

Erskine, R. W., Field, B. O.: Reversible Oxygenation. Vol. 28, pp. 1–50.

Fajans, K.: Degrees of Polarity and Mutual Polarization of Ions in the Molecules of Alkali Fluorides, SrO, and BaO. Vol. 3, pp. 88–105.

Fee, J. A.: Copper Proteins – Systems Containing the "Blue" Copper Center. Vol. 23, pp. 1–60.
Feeney, R. E., Komatsu, S. K.: The Transferrins. Vol. 1, pp. 149–206.
Felsche, J.: The Crystal Chemistry of the Rare-Earth Silicates. Vol. 13, pp. 99–197.
Ferreira, R.: Paradoxical Violations of Koopmans' Theorem, with Special Reference to the 3d Transition Elements and the Lanthanides. Vol. 31, pp. 1–21.
Fidelis, I. K., Mioduski, T.: Double-Double Effect in the Inner Transition Elements. Vol. 47, pp. 27–51.
Fournier, J. M.: Magnetic Properties of Actinide Solids. Vol. 59/60, pp. 127–196.
Fournier, J. M., Manes, L.: Actinide Solids. 5f Dependence of Physical Properties. Vol. 59/60, pp. 1–56.
Fraga, S., Valdemoro, C.: Quantum Chemical Studies on the Submolecular Structure of the Nucleic Acids. Vol. 4, pp. 1–62.
Fraústo da Silva, J. J. R., Williams, R. J. P.: The Uptake of Elements by Biological Systems. Vol. 29, pp. 67–121.
Fricke, B.: Superheavy Elements. Vol. 21, pp. 89–144.
Fuhrhop, J.-H.: The Oxidation States and Reversible Redox Reactions of Metalloporphyrins. Vol. 18, pp. 1–67.
Furlani, C., Cauletti, C.: He(I) Photoelectron Spectra of *d*-metal Compounds. Vol. 35, pp. 119–169.
Gázquez, J. L., Vela, A., Galván, M.: Fukui Function, Electronegativity and Hardness in the Kohn-Sham Theory. Vol. 66, pp. 79–98.
Gerloch, M., Harding, J. H., Woolley, R. G.: The Context and Application of Ligand Field Theory. Vol. 46, pp. 1–46.
Gillard, R. D., Mitchell, P. R.: The Absolute Configuration of Transition Metal Complexes. Vol. 7, pp. 46–86.
Gleitzer, C., Goodenough, J. B.: Mixed-Valence Iron Oxides. Vol. 61, pp. 1–76.
Gliemann, G., Yersin, H.: Spectroscopic Properties of the Quasi One-Dimensional Tetracyanoplatinate(II) Compounds. Vol. 62, pp. 87–153.
Golovina, A. P., Zorov, N. B., Runov, V. K.: Chemical Luminescence Analysis of Inorganic Substances. Vol. 47, pp. 53–119.
Green, J. C.: Gas Phase Photoelectron Spectra of *d*- and *f*-Block Organometallic Compounds. Vol. 43, pp. 37–112.
Grenier, J. C., Pouchard, M., Hagenmuller, P.: Vacancy Ordering in Oxygen-Deficient Perovskite-Related Ferrities. Vol. 47, pp. 1–25.
Griffith, J. S.: On the General Theory of Magnetic Susceptibilities of Polynuclear Transitionmetal Compounds. Vol. 10, pp. 87–126.
Gubelmann, M. H., Williams, A. F.: The Structure and Reactivity of Dioxygen Complexes of the Transition Metals. Vol. 55, pp. 1–65.
Guilard, R., Lecomte, C., Kadish, K. M.: Synthesis, Electrochemistry, and Structural Properties of Porphyrins with Metal-Carbon Single Bonds and Metal-Metal Bonds. Vol. 64, pp. 205–268.
Gütlich, P.: Spin Crossover in Iron(II)-Complexes. Vol. 44, pp. 83–195.
Gutmann, V., Mayer, U.: Thermochemistry of the Chemical Bond. Vol. 10, pp. 127–151.
Gutmann, V., Mayer, U.: Redox Properties: Changes Effected by Coordination. Vol. 15, pp. 141–166.
Gutmann, V., Mayer, H.: Application of the Functional Approach to Bond Variations under Pressure. Vol. 31, pp. 49–66.
Hall, D. I., Ling, J. H., Nyholm, R. S.: Metal Complexes of Chelating Olefin-Group V Ligands. Vol. 15, pp. 3–51.
Harnung, S. E., Schäffer, C. E.: Phase-fixed 3-Γ Symbols and Coupling Coefficients for the Point Groups. Vol. 12, pp. 201–255.
Harnung, S. E., Schäffer, C. E.: Real Irreducible Tensorial Sets and their Application to the Ligand-Field Theory. Vol. 12, pp. 257–295.
Hathaway, B. J.: The Evidence for "Out-of-the-Plane" Bonding in Axial Complexes of the Copper(II) Ion. Vol. 14, pp. 49–67.
Hathaway, B. J.: A New Look at the Stereochemistry and Electronic Properties of Complexes of the Copper(II) Ion. Vol. 57, pp. 55–118.
Hellner, E. E.: The Frameworks (Bauverbände) of the Cubic Structure Types. Vol. 37, pp. 61–140.
von Herigonte, P.: Electron Correlation in the Seventies. Vol. 12, pp. 1–47.
Hemmerich, P., Michel, H., Schug, C., Massey, V.: Scope and Limitation of Single Electron Transfer in Biology. Vol. 48, pp. 93–124.

Hider, R. C.: Siderophores Mediated Absorption of Iron. Vol. 58, pp. 25–88.
Hill, H. A. O., Röder, A., Williams, R. J. P.: The Chemical Nature and Reactivity of Cytochrome P-450. Vol. 8, pp. 123–151.
Hogenkamp, H. P. C., Sando, G. N.: The Enzymatic Reduction of Ribonucleotides. Vol. 20, pp. 23–58.
Hoffmann, D. K., Ruedenberg, K., Verkade, J. G.: Molecular Orbital Bonding Concepts in Polyatomic Molecules – A Novel Pictorial Approach. Vol. 33, pp. 57–96.
Hubert, S., Hussonnois, M., Guillaumont, R.: Measurement of Complexing Constants by Radiochemical Methods. Vol. 34, pp. 1–18.
Hudson, R. F.: Displacement Reactions and the Concept of Soft and Hard Acids and Bases. Vol. 1, pp. 221–223.
Hulliger, F.: Crystal Chemistry of Chalcogenides and Pnictides of the Transition Elements. Vol. 4, pp. 83–229.
Ibers, J. A., Pace, L. J., Martinsen, J., Hoffman, B. M.: Stacked Metal Complexes: Structures and Properties. Vol. 50, pp. 1–55.
Iqbal, Z.: Intra- und Inter-Molecular Bonding and Structure of Inorganic Pseudohalides with Triatomic Groupings. Vol. 10, pp. 25–55.
Izatt, R. M., Eatough, D. J., Christensen, J. J.: Thermodynamics of Cation-Macrocyclic Compound Interaction. Vol. 16, pp. 161–189.
Jain, V. K., Bohra, R., Mehrotra, R. C.: Structure and Bonding in Organic Derivatives of Antimony(V). Vol. 52, pp. 147–196.
Jerome-Lerutte, S.: Vibrational Spectra and Structural Properties of Complex Tetracyanides of Platinum, Palladium and Nickel. Vol. 10, pp. 153–166.
Jørgensen, C. K.: Electric Polarizability, Innocent Ligands and Spectroscopic Oxidation States. Vol. 1, pp. 234–248.
Jørgensen, C. K.: Recent Progress in Ligand Field Theory. Vol. 1, pp. 3–31.
Jørgensen, C. K.: Relations between Softness, Covalent Bonding, Ionicity and Electric Polarizability. Vol. 3, pp. 106–115.
Jørgensen, C. K.: Valence-Shell Expansion Studied by Ultra-violet Spectroscopy. Vol. 6, pp. 94–115.
Jørgensen, C. K.: The Inner Mechanism of Rare Earths Elucidated by Photo-Electron Spectra. Vol. 13, pp. 199–253.
Jørgensen, C. K.: Partly Filled Shells Constituting Anti-bonding Orbitals with Higher Ionization Energy than their Bonding Counterparts. Vol. 22, pp. 49–81.
Jørgensen, C. K.: Photo-electron Spectra of Non-metallic Solids and Consequences for Quantum Chemistry. Vol. 24, pp. 1–58.
Jørgensen, C. K.: Narrow Band Thermoluminescence (Candoluminescence) of Rare Earths in Auer Mantles. Vol. 25, pp. 1–20.
Jørgensen, C. K.: Deep-lying Valence Orbitals and Problems of Degeneracy and Intensities in Photoelectron Spectra. Vol. 30, pp. 141–192.
Jørgensen, C. K.: Predictable Quarkonium Chemistry. Vol. 34, pp. 19–38.
Jørgensen, C. K.: The Conditions for Total Symmetry Stabilizing Molecules, Atoms, Nuclei and Hadrons. Vol. 43, pp. 1–36.
Jørgensen, C. K., Reisfeld, R.: Uranyl Photophysics. Vol. 50, pp. 121–171.
O'Keeffe, M.: The Prediction and Interpretation of Bond Lengths in Crystals. Vol. 71, pp. 161–190.
O'Keeffe, M., Hyde, B. G.: An Alternative Approach to Non-Molecular Crystal Structures with Emphasis on the Arrangements of Cations. Vol. 61, pp. 77–144.
Kahn, O.: Magnetism of the Heteropolymetallic Systems. Vol. 68, pp. 89–167.
Kimura, T.: Biochemical Aspects of Iron Sulfur Linkage in None-Heme Iron Protein, with Special Reference to "Adrenodoxin". Vol. 5, pp. 1–40.
Kitagawa, T., Ozaki, Y.: Infrared and Raman Spectra of Metalloporphyrins. Vol. 64, pp. 71–114.
Kiwi, J., Kalyanasundaram, K., Grätzel, M.: Visible Light Induced Cleavage of Water into Hydrogen and Oxygen in Colloidal and Microheterogeneous Systems. Vol. 49, pp. 37–125.
Kjekshus, A., Rakke, T.: Considerations on the Valence Concept. Vol. 19, pp. 45–83.
Kjekshus, A., Rakke, T.: Geometrical Considerations on the Marcasite Type Structure. Vol. 19, pp. 85–104.
König, E.: The Nephelauxetic Effect. Calculation and Accuracy of the Interelectronic Repulsion Parameters I. Cubic High-Spin d^2, d^3, d^7 and d^8 Systems. Vol. 9, pp. 175–212.
Köpf-Maier, P., Köpf, H.: Transition and Main-Group Metal Cyclopentadienyl Complexes: Preclinical Studies on a Series of Antitumor Agents of Different Structural Type. Vol. 70, pp. 103–185.

Koppikar, D. K., Sivapullaiah, P. V., Ramakrishnan, L., Soundararajan, S.: Complexes of the Lanthanides with Neutral Oxygen Donor Ligands. Vol. 34, pp. 135–213.

Krause, R.: Synthesis of Ruthenium(II) Complexes of Aromatic Chelating Heterocycles: Towards the Design of Luminescent Compounds. Vol. 67, pp. 1–52.

Krumholz, P.: Iron(II) Diimine and Related Complexes. Vol. 9, pp. 139–174.

Kustin, K., McLeod, G. C., Gilbert, T. R., Briggs, LeB. R., 4th.: Vanadium and Other Metal Ions in the Physiological Ecology of Marine Organisms. Vol. 53, pp. 137–158.

Labarre, J. F.: Conformational Analysis in Inorganic Chemistry: Semi-Empirical Quantum Calculation vs. Experiment. Vol. 35, pp. 1–35.

Lammers, M., Follmann, H.: The Ribonucleotide Reductases: A Unique Group of Metalloenzymes Essential for Cell Proliferation. Vol. 54, pp. 27–91.

Lehn, J.-M.: Design of Organic Complexing Agents. Strategies towards Properties. Vol. 16, pp. 1–69.

Linarès, C., Louat, A., Blanchard, M.: Rare-Earth Oxygen Bonding in the $LnMO_4$Xenotime Structure. Vol. 33, pp. 179–207.

Lindskog, S.: Cobalt(II) in Metalloenzymes. A Reporter of Structure-Function Relations. Vol. 8, pp. 153–196.

Liu, A., Neilands, J. B.: Mutational Analysis of Rhodotorulic Acid Synthesis in *Rhodotorula pilimanae*. Vol. 58, pp. 97–106.

Livorness, J., Smith, T.: The Role of Manganese in Photosynthesis. Vol. 48, pp. 1–44.

Llinás, M.: Metal-Polypeptide Interactions: The Conformational State of Iron Proteins. Vol. 17, pp. 135–220.

Lucken, E. A. C.: Valence-Shell Expansion Studied by Radio-Frequency Spectroscopy. Vol. 6, pp. 1–29.

Ludi, A., Güdel, H. U.: Structural Chemistry of Polynuclear Transition Metal Cyanides. Vol. 14, pp. 1–21.

Lutz, H. D.: Bonding and Structure of Water Molecules in Solid Hydrates. Correlation of Spectroscopic and Structural Data. Vol. 69, pp. 125.

Maggiora, G. M., Ingraham, L. L.: Chlorophyll Triplet States. Vol. 2, pp. 126–159.

Magyar, B.: Salzebullioskopie III. Vol. 14, pp. 111–140.

Makovicky, E., Hyde, B. G.: Non-Commensurate (Misfit) Layer Structures. Vol. 46, pp. 101–170.

Manes, L., Benedict, U.: Structural and Thermodynamic Properties of Actinide Solids and Their Relation to Bonding. Vol. 59/60, pp. 75–125.

Mann, S.: Mineralization in Biological Systems. Vol. 54, pp. 125–174.

Mason, S. F.: The Ligand Polarization Model for the Spectra of Metal Complexes: The Dynamic Coupling Transition Probabilities. Vol. 39, pp. 43–81.

Mathey, F., Fischer, J., Nelson, J. H.: Complexing Modes of the Phosphole Moiety. Vol. 55, pp. 153–201.

Mayer, U., Gutmann, V.: Phenomenological Approach to Cation-Solvent Interactions. Vol. 12, pp. 113–140.

Mildvan, A. S., Grisham, C. M.: The Role of Divalent Cations in the Mechanism of Enzyme Catalyzed Phosphoryl and Nucleotidyl. Vol. 20, pp. 1–21.

Mingos, D. M. P., Hawes, J. C.: Complementary Spherical Electron Density Model. Vol. 63, pp. 1–63.

Mingos, D. M. P., Johnston, R. L.: Theoretical Models of Cluster Bonding. Vol. 68, pp. 29–87.

Mingos, D. M. P., Zhenyang, L.: Non-Bonding Orbitals in Co-Ordination, Hydrocarbon and Cluster Compounds. Vol. 71, pp. 1–56.

Moreau-Colin, M. L.: Electronic Spectra and Structural Properties of Complex Tetracyanides of Platinum, Palladium and Nickel. Vol. 10, pp. 167–190.

Morgan, B., Dolphin, D.: Synthesis and Structure of Biometic Porphyrins. Vol. 64, pp. 115–204.

Morris, D. F. C.: Ionic Radii and Enthalpies of Hydration of Ions. Vol. 4, pp. 63–82.

Morris, D. F. C.: An Appendix to Structure and Bonding. Vol. 4 (1968). Vol. 6, pp. 157–159.

Mortensen, O. S.: A Noncommuting-Generator Approach to Molecular Symmetry. Vol. 68, pp. 1–28.

Mortier, J. W.: Electronegativity Equalization and its Applications. Vol. 66, pp. 125–143.

Müller, A., Baran, E. J., Carter, R. O.: Vibrational Spectra of Oxo-, Thio-, and Selenometallates of Transition Elements in the Solid State. Vol. 26, pp. 81–139.

Müller, A., Diemann, E., Jørgensen, C. K.: Electronic Spectra of Tetrahedral Oxo, Thio and Seleno Complexes Formed by Elements of the Beginning of the Transition Groups. Vol. 14, pp. 23–47.

Müller, U.: Strukturchemie der Azide. Vol. 14, pp. 141–172.

Müller, W., Spirlet, J.-C.: The Preparation of High Purity Actinide Metals and Compounds. Vol. 59/60, pp. 57–73.

Mullay, J. J.: Estimation of Atomic and Group Electronegativities. Vol. 66, pp. 1–25.

Murrell, J. N.: The Potential Energy Surfaces of Polyatomic Molecules. Vol. 32, pp. 93–146.

Naegele, J. R., Ghijsen, J.: Localization and Hybridization of 5f States in the Metallic and Ionic Bond as Investigated by Photoelectron Spectroscopy. Vol. 59/60, pp. 197–262.

Nag, K., Bose, S. N.: Chemistry of Tetra- and Pentavalent Chromium. Vol. 63, pp. 153–197.

Neilands, J. B.: Naturally Occurring Non-porphyrin Iron Compounds. Vol. 1, pp. 59–108.

Neilands, J. B.: Evolution of Biological Iron Binding Centers. Vol. 11, pp. 145–170.

Neilands, J. B.: Methodology of Siderophores. Vol. 58, pp. 1–24.

Nieboer, E.: The Lanthanide Ions as Structural Probes in Biological and Model Systems. Vol. 22, pp. 1–47.

Novack, A.: Hydrogen Bonding in Solids. Correlation of Spectroscopic and Christallographic Data. Vol. 18, pp. 177–216.

Nultsch, W., Häder, D.-P.: Light Perception and Sensory Transduction in Photosynthetic Prokaryotes. Vol. 41, pp. 111–139.

Odom, J. D.: Selenium Biochemistry. Chemical and Physical Studies. Vol. 54, pp. 1–26.

Oelkrug, D.: Absorption Spectra and Ligand Field Parameters of Tetragonal 3*d*-Transition Metal Fluorides. Vol. 9, pp. 1–26.

Oosterhuis, W. T.: The Electronic State of Iron in Some Natural Iron Compounds: Determination by Mössbauer and ESR Spectroscopy. Vol. 20, pp. 59–99.

Orchin, M., Bollinger, D. M.: Hydrogen-Deuterium Exchange in Aromatic Compounds. Vol. 23, pp. 167–193.

Peacock, R. D.: The Intensities of Lanthanide $f \longleftrightarrow f$ Transitions. Vol. 22, pp. 83–122.

Penneman, R. A., Ryan, R. R., Rosenzweig, A.: Structural Systematics in Actinide Fluoride Complexes. Vol. 13, pp. 1–52.

Powell, R. C., Blasse, G.: Energy Transfer in Concentrated Systems. Vol. 42, pp. 43–96.

Que, Jr., L.: Non-Heme Iron Dioxygenases. Structure and Mechanism. Vol. 40, pp. 39–72.

Ramakrishna, V. V., Patil, S. K.: Synergic Extraction of Actinides. Vol. 56, pp. 35–90.

Raymond, K. N., Smith, W. L.: Actinide-Specific Sequestering Agents and Decontamination Applications. Vol. 43, pp. 159–186.

Reedijk, J., Fichtinger-Schepman, A. M. J., Oosterom, A. T. van, Putte, P. van de: Platinum Amine Coordination Compounds as Anti-Tumor Drugs. Molecular Aspects of the Mechanism of Action. Vol. 67, pp. 53–89.

Reinen, D.: Ligand-Field Spectroscopy and Chemical Bonding in Cr^{3+}-Containing Oxidic Solids. Vol. 6, pp. 30–51.

Reinen, D.: Kationenverteilung zweiwertiger $3d^n$-Ionen in oxidischen Spinell-, Granat- und anderen Strukturen. Vol. 7, pp. 114–154.

Reinen, D., Friebel, C.: Local and Cooperative Jahn-Teller Interactions in Model Structures. Spectroscopic and Structural Evidence. Vol. 37, pp. 1–60.

Reisfeld, R.: Spectra and Energy Transfer of Rare Earths in Inorganic Glasses. Vol. 13, pp. 53–98.

Reisfeld, R.: Radiative and Non-Radiative Transitions of Rare Earth Ions in Glasses. Vol. 22, pp. 123–175.

Reisfeld, R.: Excited States and Energy Transfer from Donor Cations to Rare Earths in the Condensed Phase. Vol. 30, pp. 65–97.

Reisfeld, R., Jørgensen, C. K.: Luminescent Solar Concentrators for Energy Conversion. Vol. 49, pp. 1–36.

Reisfeld, R., Jørgensen, C. K.: Excited States of Chromium(III) in Translucent Glass-Ceramics as Prospective Laser Materials. Vol. 69, pp. 63–96.

Russo, V. E. A., Galland, P.: Sensory Physiology of *Phycomyces Blakesleeanus*. Vol. 41, pp. 71–110.

Rüdiger, W.: Phytochrome, a Light Receptor of Plant Photomorphogenesis. Vol. 40, pp. 101–140.

Ryan, R. R., Kubas, G. J., Moody, D. C., Eller, P. G.: Structure and Bonding of Transition Metal-Sulfur Dioxide Complexes. Vol. 46, pp. 47–100.

Sadler, P. J.: The Biological Chemistry of Gold: A Metallo-Drug and Heavy-Atom Label with Variable Valency. Vol. 29, pp. 171–214.

Schäffer, C. E.: A Perturbation Representation of Weak Covalent Bonding. Vol. 5, pp. 68–95.

Schäffer, C. E.: Two Symmetry Parameterizations of the Angular-Overlap Model of the Ligand-Field. Relation to the Crystal-Field Model. Vol. 14, pp. 69–110.

Scheidt, W. R., Lee, Y. J.: Recent Advances in the Stereochemistry of Metallotetrapyrroles. Vol. 64, pp. 1–70.

Schmid, G.: Developments in Transition Metal Cluster Chemistry. The Way to Large Clusters. Vol. 62, pp. 51–85.

Schmidt, P. C.: Electronic Structure of Intermetallic B 32 Type Zintl Phases. Vol. 65, pp. 91–133.

Schmidtke, H.-H., Degen, J.: A Dynamic Ligand Field Theory for Vibronic Structures Rationalizing Electronic Spectra of Transition Metal Complex Compounds. Vol. 71, pp. 99–124.

Schneider, W.: Kinetics and Mechanism of Metalloporphyrin Formation. Vol. 23, pp. 123–166.

Schubert, K.: The Two-Correlations Model, a Valence Model for Metallic Phases. Vol. 33, pp. 139–177.

Schutte, C. J. H.: The Ab-Initio Calculation of Molecular Vibrational Frequencies and Force Constants. Vol. 9, pp. 213–263.

Schweiger, A.: Electron Nuclear Double Resonance of Transition Metal Complexes with Organic Ligands. Vol. 51, pp. 1–122.

Sen, K. D., Böhm, M. C., Schmidt, P. C.: Electronegativity of Atoms and Molecular Fragments. Vol. 66, pp. 99–123.

Shamir, J.: Polyhalogen Cations. Vol. 37, pp. 141–210.

Shannon, R. D., Vincent, H.: Relationship between Covalency, Interatomic Distances, and Magnetic Properties in Halides and Chalcogenides. Vol. 19, pp.1–43.

Shriver, D. F.: The Ambident Nature of Cyanide. Vol. 1, pp. 32–58.

Siegel, F. L.: Calcium-Binding Proteins. Vol. 17, pp. 221–268.

Simon, A.: Structure and Bonding with Alkali Metal Suboxides. Vol. 36, pp. 81–127.

Simon, W., Morf, W. E., Meier, P. Ch.: Specificity for Alkali and Alkaline Earth Cations of Synthetic and Natural Organic Complexing Agents in Membranes. Vol. 16, pp. 113–160.

Simonetta, M., Gavezzotti, A.: Extended Hückel Investigation of Reaction Mechanisms. Vol. 27, pp. 1–43.

Sinha, S. P.: Structure and Bonding in Highly Coordinated Lanthanide Complexes. Vol. 25, pp. 67–147.

Sinha, S. P.: A Systematic Correlation of the Properties of the f-Transition Metal Ions. Vol. 30, pp. 1–64.

Schmidt, W.: Physiological Bluelight Reception. Vol. 41, pp. 1–44.

Smith, D. W.: Ligand Field Splittings in Copper(II) Compounds. Vol. 12, pp. 49–112.

Smith, D. W., Williams, R. J. P.: The Spectra of Ferric Haems and Haemoproteins, Vol. 7, pp. 1–45.

Smith, D. W.: Applications of the Angular Overlap Model. Vol. 35, pp. 87–118.

Solomon, E. I., Penfield, K. W., Wilcox, D. E.: Active Sites in Copper Proteins. An Electric Structure Overview. Vol. 53, pp. 1–56.

Somorjai, G. A., Van Hove, M. A.: Adsorbed Monolayers on Solid Surfaces. Vol. 38, pp. 1–140.

Speakman, J. C.: Acid Salts of Carboxylic Acids, Crystals with some "Very Short" Hydrogen Bonds. Vol. 12, pp. 141–199.

Spiro, G., Saltman, P.: Polynuclear Complexes of Iron and their Biological Implications. Vol. 6, pp. 116–156.

Strohmeier, W.: Problem and Modell der homogenen Katalyse. Vol. 5, pp. 96–117.

Sugiura, Y., Nomoto, K.: Phytosiderophores – Structures and Properties of Mugineic Acids and Their Metal Complexes. Vol. 58, pp. 107–135.

Tam, S.-C., Williams, R. J. P.: Electrostatics and Biological Systems. Vol. 63, pp. 103–151.

Teller, R., Bau, R. G.: Crystallographic Studies of Transition Metal Hydride Complexes. Vol. 44, pp. 1–82.

Thompson, D. W.: Structure and Bonding in Inorganic Derivates of β-Diketones. Vol. 9, pp. 27–47.

Thomson, A. J., Williams, R. J. P., Reslova, S.: The Chemistry of Complexes Related to *cis*-$Pt(NH_3)_2Cl_2$. An Anti-Tumor Drug. Vol. 11, pp. 1–46.

Tofield, B. C.: The Study of Covalency by Magnetic Neutron Scattering. Vol. 21, pp. 1–87.

Trautwein, A.: Mössbauer-Spectroscopy on Heme Proteins. Vol. 20, pp. 101–167.

Tressaud, A., Dance, J.-M.: Relationships Between Structure and Low-Dimensional Magnetism in Fluorides. Vol. 52, pp. 87–146.

Tributsch, H.: Photoelectrochemical Energy Conversion Involving Transition Metal d-States and Intercalation of Layer Compounds. Vol. 49, pp. 127–175.

Truter, M. R.: Structures of Organic Complexes with Alkali Metal Ions. Vol. 16, pp. 71–111.

Umezawa, H., Takita, T.: The Bleomycins: Antitumor Copper-Binding Antibiotics. Vol. 40, pp. 73–99.

Vahrenkamp, H.: Recent Results in the Chemistry of Transition Metal Clusters with Organic Ligands. Vol. 32, pp. 1–56.

Valach, F., Koreň, B., Sivý, P., Melník, M.: Crystal Structure Non-Rigidity of Central Atoms for Mn(II), Fe(II), Fe(III), Co(II), Co(III), Ni(II), Cu(II) and Zn(II) Complexes. Vol. 55, pp. 101–151.

Wallace, W. E., Sankar, S. G., Rao, V. U. S.: Field Effects in Rare-Earth Intermetallic Compounds. Vol. 33, pp. 1–55.

Warren, K. D.: Ligand Field Theory of Metal Sandwich Complexes. Vol. 27, pp. 45–159.

Warren, K. D.: Ligand Field Theory of f-Orbital Sandwich Complexes. Vol. 33, pp. 97–137.

Warren, K. D.: Calculations of the Jahn-Teller Coupling Costants for d^x Systems in Octahedral Symmetry via the Angular Overlap Model. Vol. 57, pp. 119–145.

Watson, R. E., Perlman, M. L.: X-Ray Photoelectron Spectroscopy. Application to Metals and Alloys. Vol. 24, pp. 83–132.

Weakley, T. J. R.: Some Aspects of the Heteropolymolybdates and Heteropolytungstates. Vol. 18, pp. 131–176.

Wendin, G.: Breakdown of the One-Electron Pictures in Photoelectron Spectra. Vol. 45, pp. 1–130.

Weissbluth, M.: The Physics of Hemoglobin. Vol. 2, pp. 1–125.

Weser, U.: Chemistry and Structure of some Borate Polyol Compounds. Vol. 2, pp. 160–180.

Weser, U.: Reaction of some Transition Metals with Nucleic Acids and their Constituents. Vol. 5, pp. 41–67.

Weser, U.: Structural Aspects and Biochemical Function of Erythrocuprein. Vol. 17, pp. 1–65.

Weser, U.: Redox Reactions of Sulphur-Containing Amino-Acid Residues in Proteins and Metalloproteins, an XPS-Study. Vol. 61, pp. 145–160.

Willemse, J., Cras, J. A., Steggerda, J. J., Keijzers, C. P.: Dithiocarbamates of Transition Group Elements in "Unusual" Oxidation State. Vol. 28, pp. 83–126.

Williams, R. J. P.: The Chemistry of Lanthanide Ions in Solution and in Biological Systems. Vol. 50, pp. 79–119.

Williams, R. J. P., Hale, J. D.: The Classification of Acceptors and Donors in Inorganic Reactions. Vol. 1, pp. 249–281.

Williams, R. J. P., Hale, J. D.: Professor Sir Ronald Nyholm. Vol. 15, pp. 1 and 2.

Wilson, J. A.: A Generalized Configuration-Dependent Band Model for Lanthanide Compounds and Conditions for Interconfiguration Fluctuations. Vol. 32, pp. 57–91.

Winkler, R.: Kinetics and Mechanism of Alkali Ion Complex Formation in Solution. Vol. 10, pp. 1–24.

Wood, J. M., Brown, D. G.: The Chemistry of Vitamin B_{12}-Enzymes. Vol. 11, pp. 47–105.

Woolley, R. G.: Natural Optical Activity and the Molecular Hypothesis. Vol. 52, pp. 1–35.

Wüthrich, K.: Structural Studies of Hemes and Hemoproteins by Nuclear Magnetic Resonance Spectroscopy. Vol. 8, pp. 53–121.

Xavier, A. V., Moura, J. J. G., Moura, I.: Novel Structures in Iron-Sulfur Proteins. Vol. 43, pp. 187–213.

Zumft, W. G.: The Molecular Basis of Biological Dinitrogen Fixation. Vol. 29, pp. 1–65.

GPSR Compliance

The European Union's (EU) General Product Safety Regulation (GPSR) is a set of rules that requires consumer products to be safe and our obligations to ensure this.

If you have any concerns about our products, you can contact us on ProductSafety@springernature.com

In case Publisher is established outside the EU, the EU authorized representative is:

Springer Nature Customer Service Center GmbH
Europaplatz 3
69115 Heidelberg, Germany

Batch number: 10620881

Printed by Printforce, the Netherlands